AF815460

Theories and Theorems

(Common Theories and Laws of Physics Explained)

Authored By

Mita Thakur
British Science Association
Derby, UK

CONTENTS

CHAPTERS

Foreword

I am very pleased to have the privilege of writing a foreword for this eBook. Her eBook is explicitly not aimed at students but at everyone who is interested in physics and who wants to expand his toolkit and start speaking the language of physicists.

The eBook covers important laws, theories and theorems which are very important to know: what is the world, where and how we live, what was the past and what will be future? The content is very similar to the book "Elementary Particles and the Laws of Physics" by Richard Phillips Feynman, and Steven Weinberg, It might also remind us of our knowledge of physics. However, what's special about this eBook is that she introduces along the way all the concepts that are needed for general readers with interest in physics. The eBook is really an interesting addition to the vast collection of knowledge and because of its approach towards the general reader; this eBook is quite an achievement.

It can be useful for teenagers, and those who have many questions; because there is a gap in the physics books between high school level and the graduate level Physics students - this eBook can bridge the gap.

However, if you think that this eBook will bring them up to the level of a person having a physics degree, you got it wrong. It's a long way to a deep knowledge of important theories and there are no shortcuts for it. Thakur's eBook, and the ones that will probably follow, however might be the shortest route, to understand the reality of physics.

I recommend the teenagers, students and the curious readers to enjoy this eBook and use this as a lighter to fire their quest for the knowledge of physics.

N. Jha
B.R.A. Bihar University
India

Preface

Science book writers often explain and justify their purposes to the readers through the prefaces of their book. Here only once does the author speak to the reader with an individual voice. After that he or she vanishes from the work. In prefaces, we have the opportunity to gain direct insight into the ideas of science writers.

Waves crashing on the beach, the fluorescent jellyfish, the crystals, the sunlight, earthquake and a sneeze… many questions

Snap a photo, and you are done – static electricity at work!

This eBook focuses on how physics and its properties are used to make our everyday lives easy. The eBook originated as an attempt to collect the most interesting facts and theories of physics in such a way so as to give the reader a general picture of the basic view of phenomena; as it presents itself to the scientists. Some chapters of the eBook are simple enough to be understood by a teenager whereas others will require more concentration and further study to be completely understood, however, a layman reader will not find any difficulty in reading this eBook.

This eBook is intended for the general reader who has an interest in questioning; it aims to develop an understanding of the important concepts of physics and their applications.

The thought that I should write this eBook has been with me since the moment I started to paint: the question arises.

Why is the sky blue? (p. 85).

My father was a physicist and he told me simply why the sky is blue? And what makes the sunset red? That was quite interesting for me and I decided to study physics. I realized how beautiful and elegant physics is!

How small acts have a big concept and a large things have a small theory!

I think this eBook will provide an interest about Physics to the readers, which challenges our imaginations with concepts like relativity and string theory, and it leads to great discoveries, like computers and lasers, that lead to technologies which change our lives—from healing joints, to curing cancer and to developing sustainable energy solutions.

I hope that this eBook will help a new generation of scientists and users of new technology to solve complicated problems with greater understanding; and belief that the publication of this eBook will enhance the spread of ideas that tickle the mind.

I want to express my special thanks to my teachers, my tutors and my guide for their extensive help and support.

CONFLICT OF INTEREST

The author confirms that this eBook contents have no conflict of interest.

Mita Thakur
British Science Association
Derby
UK
E-mail: Mita.thakur@gmail.com

Acknowledgements

The eBook is aimed for the general audience; therefore authorities have not been quoted in the text. Many original papers and books have been referred to during the assembly of this volume and it would be wrong to present it without acknowledging their valuable assistance.

I would like to express my gratitude to the many people who provided support, discussed various points, read, wrote, offered feedback, allowed me to quote their remarks and assisted me in the editing and proofreading.

To my mother the Philosopher; I can barely find words to express all the wisdom, love and support she has given to me.

I owe my deepest gratitude to my parents-in-law.

Above all I want to thank my husband and my two sons, who supported me all the time.

To my husband, who has good grasp in physics, and he guided me throughout.

To my elder son, who is a doctor with great passion for physics, he offered comments and gave me his insightful feedback.

To my younger son, who is a student of international affairs and has great interest in books, assisted me in the editing and proofreading. He also developed his interest in physics while editing my eBook.

I would like to thank my publisher for enabling me to publish this eBook.

Last and not the least: I beg forgiveness of all those who have been with me over the course of the years and whose names I have failed to mention.

Introduction

The things which we consider individually in science that appear in multiple forms with a multitude of aspects. At the seashore when we look at the sea, we notice: the water, the waves breaking, the foam, the splashing motion of the water, the sound, the air, the winds and the clouds, the sun and the blue sky and light; the sand and rocks of different hardnesses, colours and textures. There are animals and seaweeds, hunger and disease, happiness and thought and the observer on the beach.

Nature has a similar variety of things and influences. It is always complicated.

Curiosity demands answers to questions?

- Is the sand a mixture of very tiny stones?
- Is the moon a great rock?
- Why are there different kinds of sounds?
- How many different colours are there?

And so on….

We try to put things together and try to understand this multitude of aspects.

A few hundred years ago, a method was devised to find partial/possible answers to such questions.

Observation, experiment and reasoning; that we call the Scientific method.

To limit ourselves to a bare description of our basic views of phenomena - is called fundamental physics, or fundamental ideas which have arisen from the application of the scientific method.

The world is like a chess game played by the God, and we are observers of the game. We do not know what the rules of the game are but we are allowed to watch the play.

As already stated, the eBook is a collection of theories and laws of Physics explained in simple language, meant for the general public with an interest in Physics.

CHAPTER 1

Terminology

Abstract: Terminology is the study of terms and their uses. These are analysed from the point of view of their origin, formal structure, their meanings and functional features. Terms are used to denote concepts. This chapter presents the technical terms of science which have very specific meanings. It focuses only on those terms which give particular attention to the eBook content.

Keywords: Atomic physics analyse, attention, classical mechanics, concept, content, description, fact, focus, formal structure, functional features, guess, hypothesis, law, meanings, nature, origin, particular, phenomenon, science, set of ideas, specific, technical, terminology, terms, theorem, theory.

FACT

A thing that is known or proved to be true. It is a truth by actual observations or experiences; something, that actually exists.

PRINCIPLE

A principle is a rule or law from which others are derived. It is an idea that forms the basis of something which explains how something works or why something happens.

LAW

A law describes a generalized pattern in nature that is supported by scientific evidence and repeated experiments. The laws of science are statements which describe, predict and explain why ranges of phenomenon occur in nature.

HYPOTHESIS

A hypothesis is a reasonable guess based on what you know or observe. This is an idea that needs to go through a number of testings before it turns into a theory.

Mita Thakur

THEORY

A theory is a set of ideas used to explain why something is true. It summarizes a hypothesis or a group of hypothesis which is supported with repeated testings.

THEOREM

A theorem is a result that can be proved to be true from a set of ideas. It is a formula or statement that can be proved from other formulas or statements.

A terminology is a science whose aim is to study terms for specialised field. Terminology allows the collection, description and presentation of terms.

A physical law signifies four different concepts:

1. Objective pattern or natural regularity
2. Formula claiming to represent an objective pattern
3. Law-based rule (uniform procedure)
4. Principle concerning with previous.

The basic laws of physics fall into two categories:

- Classical physics deals with the observable world (classical mechanics),
- Atomic physics deals with the interactions between elementary and sub atomic particles (quantum mechanics). It is the branch of physics concerned with the structure of the atom and the characteristics of subatomic particles.

Classical Physics does not make use of quantum mechanics or the theory of relativity.

Newtonian mechanics, Thermodynamics, and Maxwell's theory of electromagnetism are all examples of classical physics.

CHAPTER 2

Laws, Theories and Theorems

Abstract: Physics is the scientific study of matter and energy. It ranges from subatomic particles to stars and entire galaxies. In experimental science, physics employs the scientific method to formulate and tests the facts based on observation of the natural world. A scientific principles, laws, theories and theorems describe the behaviour of how something occurs, why something works and the related method of mathematical reasoning. These are based upon observations and repeated experimentations. They can be disproved, modified or accommodate new discoveries and also they make predictions for future experiments and observations. This chapter presents the lists of important laws, theories and theorems of physics which describe, predict, and explain why, a range of phenomena behave as they appear to be in nature. This chapter also provides the basis upon which all of physics is based.

Keywords: Buoyancy, describe, discoveries, earth science, electric current, energy, experiments, fluid mechanics, galaxy, laws, length elements, loop, magnetic field, mass, matter, nature, permeability, phenomena, physics, predict, predictions, pressure, range of phenomenon, range, reasoning, scientific study, theorems, theories, volume, weight.

AMPERE'S LAW (1836)

"At any point a magnetic field induced by an electric current is directly proportional to the product of the current intensity and the length of the current conductor; inversely proportional to the square of the distance between the point and the conductor, and perpendicular to the plane joining the point and the conductor".

This is a useful law that relates to the net magnetic field along a closed loop to the electric current passing through the loop. This has been modified by Maxwell's equations.

[In fluid mechanics and earth science, permeability is the ability of a substance to allow another substance to pass through it].

Mita Thakur

ARCHIMEDES' PRINCIPLE

"The weight of the liquid displaced by a floating body is equal to the weight of a body".

This principal applies to bodies submerged in all types of substances.

- If the weight of displaced water is less than the weight of the object, the object will sink.
- If the weight of the displaced water equals the weight of the object, the object will float.

[Buoyancy is an upward force applied by a fluid against the weight of an immersed object. It is the tendency of an object to float in a liquid or rise in air or gas…].

Example:

1. When we enter into a bath tub filled with a certain level of water, it rises based on the volume of water the weight of that the person's body displaces.

   The buoyant force that acts on a body that is partially or totally immersed in a fluid is the same as the weight of the fluid that the body displaces. The weight of an object acts downward, and the buoyant force provided by the displaced fluid acts upward. If these two forces are equal, the object floats.

   The mass of a substance per unit of volume is known as *density*. If object's density is higher than water's density, the object will sink in water.

2. A ship floating on the water displaces an amount of water weight equal to its own weight. If it displaces lesser water than its own weight, it sinks. It indicates how much load is allowed at different types of water.

AVOGADRO'S HYPOTHESIS (1811)

Equal volume of all gasses contains equal numbers of molecules at the same pressure and temperature.

PV = nRT, where P is the pressure, V is the volume, n is the number of moles of s, R is the universal gas constant and T is the temperature.

This hypothesis is true only for ideal gasses.

For *e.g.*

Lungs expand when filling with air and contracts with exhaling the air.

The weight of helium filled balloon is lesser than an identical air filled balloon.

(Both balloons contain the same number of molecules; helium atoms have lesser mass than oxygen or nitrogen molecules in the air, so the helium balloon is lighter than air filled balloon...).

[Avogadro's number is the number of particles found in one mole of a substance. It is approximately 6.022×10^{23} particles per mole]. It is also known as Avogadro's constant – one of the fundamental constants of chemistry.

AMONTON'S LAW

"In an ideal gas, at constant volume and the number of moles, the pressure of the gas is directly proportional to the absolute temperature".

Number of moles - amount of gas

Example:

- The gauge pressure in a steel belted tire of a car, when moving on hot asphalt, will be higher than that when car is in rest.
- If an aerosol can is thrown into fire it may explode.

Boyle's law, Charles' law, Avogadro's law and Amonton's law are given under certain conditions so, directly combining them will not work. The ideal gas law (or equation) is the combination of the properties of these three simple gas laws.

BOHR'S THEORY (1913)

"The electrons circulate around the nucleus in discrete, stable orbits with different energy level".

This theory is an early model of atomic structure; was first to explain the atomic spectrum of the hydrogen atom. Main points of the Bohr Model include:

1. Electrons revolve around the nucleus in orbits; have a definite size and energy.

2. The energy of the orbit is related to its size. The lowest energy is found in the smallest orbit.

3. Radiation is absorbed or emitted when an electron moves from one orbit to another.

This is a theory of atomic structure that combined Rutherford's model with the quantum theory.

BELL'S THEOREM (1964)

"No local hidden variable theory can make predictions in agreement with those of quantum mechanics".

BRANS-DICKS THEORY (1961)

A modification of the general theory of relativity that combined tensor and scalar qualities, allows the gravitational constant to vary with time, which is similar to Mach principle.

BREWSTER'S LAW (1811)

"The index of refraction for a material is equal to the tangent of the polarizing angle for the material".

It says when unpolarised light is incident at the angle of polarisation (Brewster's angle), the light which is reflected from the surface is then perfectly polarized.

BETZ'S LAW (1919)

"No turbine can capture more than 59.3 percent of the potential energy in wind".

Betz law is a theory which calculates the way to how the maximum energy could be derived from a wind turbine.

Under ideal condition the maximum energy that can be extracted from the wind is called the power coefficient, which is a ratio between the amounts of energy that can be extracted by a wind turbine to the total energy in the wind.

This law is derived from the principle of conservation of mass and momentum of the air stream flowing through an actuator disk which extracts energy from the wind (Fig. **1**).

Windmill

Fig. (1). Windmill – converting kinetic energy into electrical energy.

In the windmill, the wind turbine converts the kinetic energy of the wind into electrical energy.

BERNOULLI'S PRINCIPLE (1738)

"At any point in a pipe through which a fluid is flowing; the sum of the pressure energy, the kinetic energy and the potential energy of the fluid is constant".

This principle can be applied on various type of fluid flow and is also denoted as Bernoulli's equation. It can be derived from the principle of conservation of energy.

For *e.g.*

As pressure is lowest when the velocity of flow is greatest and *vice versa*; flow is slowest through widest parts and fastest through the thinnest parts of a tube.

Example: Aerofoil, Bunsen burner, Hydrofoil boat, Insecticide spray, Carburettor

Air Flight

In an airplane wing, the top wing is to a certain degree curved and the bottom is totally flat.

In the sky, air travels across both the top and the bottom parts. Differently designed parts allow air to move slower on bottom, which creates more pressure on bottom and allow air to move faster on top, that creates less pressure on top; so it creates lift, and allows plane to fly.

BIOT-SAVART LAW (1820)

"The magnetic induction near a long, straight conductor, as wire, is inversely proportional to the distance from the conductor and directly proportional to the intensity of the current in the conductor".

This law is related to magnetic field to the current which are their source. It gives a method to calculate the magnetic field produced by an arbitrary current distribution. This law is similar to the Coulomb's law that relates the electric fields to the point charges, which are their sources. It can be derived from Ampere's law.

BOYLE'S LAW (1622)

The volume of a given mass at constant temperature is inversely proportional to its pressure.

i.e.

$P\ V = constant$

where, V is the volume and P is the pressure. This is true only for an ideal gas.

For *e.g.*

- In scuba drive, the bubbles exhaled by the driver increases as he approaches to the surface of the ocean?

  The pressure exerted by the weight of water, decreases with the decreasing depth and the volume of the bubble and increases as they rise.

- Deep sea fish die when brought to the surface?

  The volume of the gas in their bodies increases and pops cells, bladders and membranes.

- In bicycle pump, the air in the cylinder is compressed when a piston is pushed into the cylinder?

According to Boyle's Law, the air pressure inside the cylinder will increase; this causes the air pressure in the cylinder to become higher than the pressure inside the tire. Therefore, the air can flow into the tire.

BRAGG'S LAW (1912)

When a beam of X-ray strikes a crystal surface, its angle of incidence will reflect back with a same angle of scattering; and the path difference is equal to an integer number of wavelength a constructive interference will occur.

The equation of the law explains why the faces of crystals appear to reflect (diffract) X-ray beam at certain angles of incidence.

In 1953, X-ray crystallography was used to find the double helix structure of DNA. Now there are many other ways to see inside the objects.

- Computed Tomography (CT) - It uses many X-ray slices at different angles to build up three dimensional images inside the body.
- Ultrasound - Sound waves are uses to locate different organs inside the body. It also is used to monitor the growth of unborn babies.
- Magnetic Resonance Imaging 9mri0 – It uses powerful magnet to measure the water content of different types of body tissue.
- Positron Emission Tomography (PET) - It detects radioactive tracers that are injected into the body.
- Electroencephalography (EEG) - It detects electrical activity in the brain.

Physics helps to develop most powerful diagnostic tools in medicine.

BROWNIAN MOVEMENT (1827)

"The random motion of particles suspended in a liquid or gas is caused by their collision of the particles with the molecules of the medium".

Application

- In X-ray diffraction
- In Material detection

CARNOT'S THEOREM

"No engine operating between two given temperatures can be more efficient than a reversible engine". *i.e.* The efficiency of an engine using irreversible processes cannot be greater than the efficiency of an engine using reversible process that is working between the same temperature.

CHARLES' LAW (1787)

"At constant pressure, the volume of an ideal gas is directly proportional to the absolute temperature".

For *e.g.*

Football

On a winter day a football is inflated in the home and it shrinks when it is taken to outside. This is because, according to Charles' law, a gas takes up more space when it is warm.

Automobile Tire

The automobile owner's manual mentions that tire pressure should always be measured cold. This is because driving heats up the tires. Charles' law says that gas occupies more space when heated but a tire does not expand very much, so tire pressure to be higher when warm and it will give the false impression for too much air in the tire.

Hot-Air Balloon

When the air in a balloon is heated at atmospheric pressure, its temperature will increase. According to Charles' Law, the volume of gas in the balloon will increase when its temperature increases. The weight of balloon is less than the weight of cold air, so the hot air balloon rises. When the gas in the balloon cools, it returns to the ground.

Cakes and Bread

When warm yeast and sugar are mixed, it produces carbon dioxide gas. As it warms, the carbon dioxide begins to expand and the bread begins to rise. According to the law, the heat of the oven kills the yeast so carbon dioxide stops to form and bread stops to rise.

[The ideal gas law is the combination of all simple gas laws: Boyle's law, Charles' law and Avogadro's law…..].

CAUSALITY PRINCIPLE

"Distinct causes have distinct effects".

This is the principle of relationship between cause and effect. It is particularly useful when combined with the speed of light in a vacuum. It is used to analysed the results of scattering experiments and in optics.

This principle can be understood in both directions: equal effects have equal causes.

For *e.g.*

A pinpricking causes pain.

When brain damages it cause mental illness.

CHERENKOV RADIATION (1934)

"Electromagnetic radiation, a bluish light, emitted by a beam of high energy charged particles passing through a transparent medium at a speed greater than the speed of light in that medium".

Cerenkov radiation is a bluish light frequently seen in nuclear reactors where water is used as a moderator. This radiation only occurs in a medium such as air or water, because the speed of light in the medium is slower than the speed of light in a vacuum. This type of radiation is very useful in the particle physics because of it provides a way to detect high energy particle.

Marie and Pierre Curie predicted it in 1880s. It was studied in details by Pavel Cerenkov in 1930s, and later it was explained by Frank and Tamm. The three of them won the Noble Prize in 1958 for their work.

Cerenkov radiation became very important for the studies of high energy cosmic rays and high energy gamma rays. He studied the blue light appearing when the radioactive objects were placed near water and other transparent materials. Now we study that any electrically charged particle (like electron) which travels with high energy through water, air, or any other transparent medium, will release bluish light. This light move outside from the particle at a particular angle to the particle's motion.

CASIMIR EFFECT (1948)

"A small attractive force, acts between two macroscopic conducting surfaces in a volume, only contains an electromagnetic field".

Basically, it shows the attraction between two objects if they come within, 1/5000 of an inch to each other.

It is related to the energy inherent in the quantum vacuum.

The Casimir Effect is one of the most interesting and potentially powerful phenomena known to modern science.

In a vacuum, if two mirrors are placed facing each other; some waves will tend to fit and bouncing back and forth, while others do not. As mirrors come closer to each other, longer waves will not fit between them because the total amount of energy between plates will becomes less than the amount of energy in another place in the vacuum; so mirror will attract each other. This effect of attractive force is called Casimir Effect.

CENTRIFUGAL PSEUDO FORCE

"An outward force, equal and opposite to the centripetal force, experienced by a body moving along a curved path and appearing the body to push outward".

This is an idea of a force pushing the object out of the centre of the circle. It is called pseudo because it is the force just like the Newton's third law of kind of force (action is real and reaction is virtual). Pseudo force is not real but it makes to feel present.

[Centripetal force is the inward force on a body moving in a curved path around another body…..].

For *e.g.*

If a ball is tied on the end of a string and rotated in a circular path, the ball tends to go away from centre of circle which is formed by rotating ball in a circular path. This is centrifugal force. (The string exerts centripetal force on the ball and causes it to follow a curved path).

Highway curves are tilted to prevent centrifugal force from forcing cars outward off the road (The effect of centrifugal force can be controlled and utilized for useful purposes).

COMPLEMENTARITY PRINCIPLE (1928 BY NIELS BOHR)

"A given system cannot show both wave-like and particle-like behaviours at the same time".

It is a fundamental principle of quantum mechanics.

Bohr has tried to apply this principle to the other fields like psychology for mind and body problems such as parallelism, interactionism, idealism, *etc.*

After certain experiments, this principle explains the wave-particle duality phenomenon. It gives a complete description of a phenomenon in microphysics.

COMPTON EFFECT (1923)

"The collision between a photon and a particle results in an increase in the kinetic energy of the particle and a corresponding increase in the wavelength of the photon".

This effect is about the result of high energy photon's collision with a target, which releases loosely bound electrons from the outer shell of an atom or molecule. The scattered radiation experiences a wavelength shift which cannot be explained by classical wave theory; therefore, it was supported with Einstein's photon theory. This discovery helped to establish that the electromagnetic radiation has a dual nature: as a wave and as a particle.

- In diagnostic radiology, the Compton Effect plays both positive and negative roles.

CONSTANCY PRINCIPLE (1922)

"Stability is the end goal for mental process".

In psychology, this theory states that all mental processes have a tendency toward equilibrium and stability and it is called Constancy Law. This principle gives a point of view which was widely accepted in the sciences of the 19th century.

For *e.g.* The Principle of Constancy of Light speed.

The speed of light in empty space (vacuum) is the same for all non-accelerated observers, in spite of the speed of light source or the observer.

The 'constancy' of physics is one of the most admired principles in science - but the scientists say that the 'laws' may work differently in different places.

CONSERVATION LAWS

"The physical properties remain constant throughout the various processes that occur in the physical world".

There are six conservation laws are in physics.

These have certain conditions that system should close and is not affected by anything outside it. These laws of conservation make possible to predict the macroscopic behaviour of a system without considering the microscopic details of a physical process or chemical process.

Baryon is a composite subatomic particle made up of three quarks. Baryons and mesons are from hadron family, which are the quark based particles.

Leptons and quarks are the basic building blocks of matter, *i.e.*, they are seen as the "elementary particles".

1. Conservation of mass-energy

   The total energy of a closed system remains constant.

2. Conservation of electric charge

   The total electric charge of a closed system remains constant.

3. Conservation of linear momentum

   The total linear momentum of a closed system remains constant.

4. Conservation of angular momentum

   The total angular momentum of a closed system remains constant.

5. Conservation of Baryon number

   In any reaction, the total number of baryons must remain constant.

   [Baryons are hadrons (composite particles made of quarks) composed of any three quarks].

6. Conservation of Lepton number.

   In any reaction, the total number of Lepton must remain constant.

Conservation of baryon number and the conservation of lepton number are known as the strong conservation laws. Specific quantum numbers are set to be the different fundamental particles and other conservation laws are associated with those quantum numbers.

Conservation law is a principle which states that a quantity or property remains constant during and after an interaction or process. This law can be considered to be the most fundamental principle of mechanics.

For *e.g.*

When a car engine burns fuel, it converts the fuel's energy into chemical energy to make car move.

CONTINUITY EQUATION

A mathematical equation, derived from conservation principle. It says all fluid particles that flow into any fluid region must flow out. This equation is applicable when there is some quantity which can flow or move *i.e.* a moving fluid (conservation of mass) or electrical charge (conservation of charge).

A fluid's motion must be move in such a way that its mass is conserved. Common applications of this law are pipes, tubes and ducts with flowing fluids or gasses, rivers, power plants, *etc.* This principle is derived from the fact that mass is always conserved in fluid system despite of the pipeline complexity or direction of flow.

COPERNICAN PRINCIPLE

"It is an idea that the Sun, and not the Earth, is at the centre of the universe".

The Copernican principle is one of the primary philosophical foundations of the science of astronomy. It is not a physical law in and of itself, but it does require using any hypothesis which requires us to live in a special place of the universe. This law represents a fundamental philosophical change with humanity's role in the universe in scientific terms.

CORIOLIS FORCE (1835)

"A pseudo force existing in a frame rotates with constant angular velocity to a reference frame".

It acts on a body moving in the rotating frame to deflect its motion sideways. The Coriolis force is the force that an inexperienced observer thinks is needed to push the air eastwards.

It is a imaginary force which is used to explain a deflection in the path of a body moving in latitude relative to the earth when observer from earth.

Example:

Let us rotate a turn table of a gramophone. With a piece of chalk and a ruler, when we draw a straight line quickly from the centre to the circumference and then we stop the turn table, it is seen that the chalk line is not in a straight line, it is a curve. An apparent force causes the deflection in a sense opposite to that of the rotation of record. This apparent force is called Coriolis force. This force varies with the wind speed and latitude. It is maximum at the poles and near the equator.

The main cause of Coriolis effect is the earth's rotation: as earth spins in a counter-clockwise on its axis, anything flying or flowing over distance above its surface is deflected [*e.g.* winds, aeroplanes, missiles…].

Most common examples of this effect in action are seen through the deflection of winds on earth.

CORRESPONDENCE PRINCIPLE (1920)

"In the limit of large quantum numbers, quantum mechanics must accurately reproduce the classical laws of physics".

This law was first proposed as a set of rules for understanding the spectra of simple atoms and molecules.

This principle can be applied to any system that reduces to another in a limiting case. For example Einstein's theory of special relativity reproduces the predictions of classical mechanics for objects at velocities much lesser than the speed of light.

The rules of quantum mechanics are very successful in describing microscopic objects, atoms, and elementary particles in the mesoscopic world.

[The mesoscopic world is in between the microscopic and macroscopic world….].

The macroscopic world contains the things which we can see with our naked eyes. The microscopic world contains the building blocks of matter, atoms and molecules.

COULOMB'S LAW (1785)

"The force created between two charged particles is directly proportional to the product of their charges and inversely proportional to the square of the distance between them".

This law describes the electrostatic interaction between two electrically charged particles. As distance increases, the forces and electric fields decrease. It is very useful to understand and interpret the many chemical phenomena; such as an atom is a collection of electrical charges, positively charged protons, and negatively charged electrons. Coulomb's Law is one of the basic laws of electricity in physics also known as Coulomb's inverse square law.

The physics axioms of Coulomb's Law:

"Nothing can be created or destroyed and you cannot get something for nothing".

CURIE'S LAW

"The magnetic susceptibility of a paramagnetic substance is approximately directly proportional to an applied magnetic field and inversely proportional to its absolute temperature".

$M = C \times (B / T)$

where,

M is the magnetisation, B is the magnetic field (measured in teslas), T is the absolute temperature (measured in kelvins), C is a material- specific curie constant.

[The tesla (symbol T) is the SI derived unit of magnetic flux density commonly denoted as B]

Magnetic susceptibility is a measure of the magnetization of a material in response to an applied magnetic field. If a material is heated, the magnetization will be inversely proportional to its temperature.

Magnetic thermometers are based on the Curie's law of magnetism. These thermometers are used to measure very low temperatures (<1K) by measuring the magnetic susceptibility of a paramagnetic substance.

Pierre Curie investigated ferromagnetism, paramagnetism, and diamagnetism and discovered the effects of temperature on paramagnetism which is now known as Curie's law.

CURIE-WEISS LAW

"The magnetic susceptibility of a paramagnetic substance is inversely proportional to the difference between its temperature and curie point".

This law describes the magnetic susceptibility of a ferromagnet in the paramagnetic region above the Curie temperature.

Electric susceptibility is the quantitative measure of an electric field applied to a dielectric material causes polarization.

Curie point is the temperature at which a ferromagnetic substance looses its ferromagnetism and becomes only paramagnetic.

- Paramagnetic materials: *e.g.* platinum
- Ferromagnetic materials: *e.g.* iron

Paramagnetic substances have a small, positive susceptibility to magnetic fields.

Ferromagnetic substances are strongly attracted by a magnet; it has a large, positive susceptibility to an external magnetic field. The electrical properties of ferromagnetic substance are related with Curie-Weiss law.

DALTON'S LAW OF PARTIAL PRESSURE

"Total pressure of a gas in a container is the sum of the partial pressures of the individual gasses in the container".

Dalton's law of partial pressure is used to determine the individual pressures of each gas in a mixture of gasses.

For *e.g.* Air

Air is the mixture of nitrogen, oxygen, CO_2, H_2O, *etc*. Each of these gasses has its own partial pressure and the total makes the atmospheric pressure of that location.

The partial pressure of the water in this mixture can be translated into the humidity of the air.

(The partial pressure of a gas is a measure of the concentration of a gas in a mixture of gases. This is the pressure that would be exerted by one of the gases in a mixture if it occupied the same volume on its own…..).

DULONG-PETIT LAW (1819)

"The gram atomic heat capacity of an element is a constant and is same for all solid elements, about six calories per gram atom".

In 1819, Dulong and Petit gives their law on relation between the average specific heat of an element and its atomic weight.

Mass specific heat capacity is the amount of heat required to raise the temperature of a body per unit of mass. It is also known as specific heat or specific heat capacity. It was modified to apply only on metallic elements. After sometimes, it was measured on low temperature which showed that the heat capacity of all solid tends to become zero at low temperature. Now, this law is used only as an approximation at intermediate high temperature.

[Atomic heat capacity is the heat capacity of a gram-atomic weight of an element…..].

DOPPLER EFFECT (1842)

"The apparent change in the observed frequency of a wave for an observer is as a result of relative motion between the source and the observer".

Waves emitted by a moving body received by an observer will be blueshifted (*i.e.* compressed) when approaching; and redshifted (*i.e.* elongated) when receding. It occurs both in sound and electromagnetic phenomena.

Example:

- An ambulance
- A police car

[When an ambulance or police car goes past, its siren is high pitched as it comes towards an observer; and low pitched as it moves from an observer. This effect, where there is a change in frequency and wavelength, is called Doppler effect….].

Redshift and blueshift are related to light, on an electromagnetic spectrum; different colours of light have different frequencies and different energies. Red light has lower energy and a lower frequency, while blue light has a higher energy and a higher frequency but shorter wavelength.

Colour is created on the basis of the wavelength of the light – In the visible spectrum, red has the longest wavelength where blue (or violet) has the shortest. If a body is moving away from you then its wavelength appears longer (or red shifted) and if a body is moving toward you then its wavelength appears shorter (or blue shifted).

EINSTEIN FIELD EQUATIONS

These are a Set of equations, which describes the gravitational effects produced by a given mass in general relativity.

General relativity explains gravity as the curvature of space-time. These equations describe the properties of a gravitational field surrounding a given mass; how an object curves space and how the curvature, in turn, stretches matter in three directions: north-south, east-west and up-down.

Einstein's Mass-Energy equation

$E = mc^2$

where, E is the energy, m is the mass and c (a universal constant) is the speed of light.

[This is the most powerful formula in physics which shows the way to more ethical power generation….].

Energy is equal to mass multiplied by the speed of light squared. This equation is about special relativity. It presents the concept that energy possesses mass and mass became a way to measure the total energy present in an object. Object always has energy of motion and it is always going with the speed of light. It is a basic equation of nuclear physics.

The reason, **Why the speed of light is squared?**

C is the speed of light, which is a huge number by itself (about 670 million miles per hour or 186,282 miles per second) and its square is there for enormous, so there is lots of energy condensed into matter.

[The square of the speed of light is approx... 448,900,000,000,000,000 in units of mph]

In this equation, energy is put equal to mass, multiplied by the square of the velocity of light, showed that very small amounts of mass may be converted into a very large amount of energy and *vice versa*. In fact, the mass and energy are in equivalent. This was demonstrated with experiment by Cockcroft and Walton in 1932.

The equivalence of mass and energy is the most important development of the Special theory of relativity. Special relativity is limited to objects that are moving at constant speed in a straight line, which is called inertial motion.

EQUIVALENCE PRINCIPLE

"At any local (small) region in space-time it is possible to formulate the equations keeping physical laws such that the effect of gravitation can be neglected".

The Equivalence principle states the equivalence between the inertial and gravitational forces of acceleration. It is used to derive the important results without solving the full equation of general relativity. The whole nature works on this principle. There is a constant pressure in our environment and on every environment in the universe, and there is a constant force which equalizes the inner pressure with the outer pressure to come to a balance state, so we come to a state of rest or comfort. This is the simple law functioning all the time. This theory began with Galileo's experiment of dropping ball from Tower of Pisa.

The views of Aristotle, Galileo and Einstein on equivalence principle:

Aristotle: A falling body depends on its weight.

Galileo: The speed of falling a body is proportional to its density not its weight.

Einstein: The gravitational field has only a relative existence.

In Equivalence principle, the weak equivalence principle is known as universality of free fall or Galilean equivalence principle.

The reason for why the free fall body from the roof of a house (in the immediate surroundings), there is no gravitational field exists:

Reasons

1. Because of the equivalence between gravitational and inertial mass, a freely falling observer won't feel his own weight, nor any effect of gravity.

2. Gravity can be transformed away.

3. Under free fall, the only force acting upon a body is the force of gravity. In fact, there is also the force of friction from the air (air resistance) that opposes the force of gravity. The force of gravity is the main acting force, air resistance acts in the opposite direction.

FARADAY'S LAW (1821)

Faraday's law is a fundamental relationship which comes from Maxwell's equation:

1. Faraday's law of induction:

It is a basic law of electromagnetism which describes how a magnetic field will interact with an electric circuit to produce an electromotive force (e.m.f).. This is called electromagnetic induction.

2. Faraday's law of electrolysis

(i) The amount of chemical change produced is proportional to the quantity of electricity passed.

(ii) The amount of chemical change produced in different substances by fixed quantity of electricity is proportional to the electrochemical equivalent of the substance.

[Magnetic flux (Φ_m) is a measure of quantity of magnetism, paying attention to the strength and the extent of a magnetic field…].

[Electromagnetic force (emf) is linked with electric and magnetic field. This force holds electron and proton together in atoms, which hold atoms together to make molecules…..].

Ampere's law relates the magnetic fields to the electric current source. Ampere's experiment with forces on wires and Faraday's law of induction were later unified by Maxwell.

Technological Applications

Transformer – transforming the level of A.C. voltage and current very easily and also very useful to transmit A.C. electricity over long distances without much loss of power. At the generating station, voltage is increased to by using step-up transformers. The distribution transformer is used at the distribution level where voltage tends to be lower and the voltage is decreased to the 230 V (A.C. Mains) by using step-down transformers for operating the household appliances, industrial machines *etc*.

e.g. Electrical Generators, Electromagnetic Flow Meters, Musical Instruments, Induction Cookers, *etc*.

Induction cooker – The burner has very powerful and high frequency electromagnet, which is generated by advanced electronics in the burner under the unit's ceramic surface.

A.C. flowing through the coil of copper wire, placed below a cooking container produces an alternating magnetic field; that magnetic field induces a current in the electrically conductive container, which produces heat in it by Joule's heating effect. Induction cookers are very fast, have better thermal efficiency and offer instantaneous control of cooking energy.

Musical instruments - like electric guitar and electric violin, they have a pick up advice attached to them. It consists of a fine enamelled copper wire wound on a magnet. When the metal strings of guitar are strung, the vibrating string cuts the magnetic flux of magnet, which links the coil due to that electric current is induced in the pick-up's coil. It is modulated by the mechanical vibration of the string; this electrical signal is then amplified and recorded by suitable devices. The phenomena of electromagnetic induction is also used in the instruments and machines e.g. induction motor, induction sealing, electrical generator, electro-magnetic flow meter, credit cards (Fig. **2**), audio-video tapes, hall effect meters, Faraday disk, *etc*.

Fig. (2). Credit card.

FERMAT'S PRINCIPLE

"Path taken by a light ray in moving from one point to another point will be the path requires the least time".

It is also called the principle of least time.

Or

"Light travels through the path in which it can reach the destination in least time"

Sometimes it is recognized as the definition of a ray of light.

Example: Magnetic spectrometers

GAUSS' LAW

"The net flux of an electric field through a closed surface is proportional to the enclosed electric charge".

This law is related to electric flux for closed surface.

- Gauss' law is the first of the four Maxwell equations which is the sum of all the electromagnetic theory.

- Gauss' law is an alternative to Coulomb's law for calculating the electric field due to a given distribution of charges.

Gauss' flux theorem is also known as Gauss's law, which relates the distribution of electric charge to the resulting electric field.

Hall Effect (1879)

"The production of an e.m.f. within a conductor or semiconductor, through which a current is flowing, a magnetic field is applied in a direction perpendicular to that of the flow of current".

The Hall effect can be used to measure magnetic fields with a hall probe.

Applications

Hall effect sensor - it is a device which is activated by an external magnetic field. It is useful for proximity switching, positioning, speed detection and current sensing applications.

Hooke's Law

"The stress applied to any solid is proportional to the strain it produces within the elastic limit for that solid".

It gives the relationship between the forces applied to a string and its elasticity. When an elastic object is stretched, the increased length is called Extension. The extension of an elastic body is directly proportional to the force applied on it.

$$F = k \times e$$

where, F is the force in Newton (N), k is the spring constant (Newton per meter N/m), e is the extension in meters (m).

[This equation works as long as the elastic limit is not exceeded. If a spring is stretched too much, it will not return to its original length when the load is removed].

It is an example of first law of thermodynamics. Hooke's law only works in a limited frame of reference; its periodic function is just like a wave.

Hooke's discovered the law of elasticity. He also built and used microscopes, which he used to study organisms. He introduced the term 'cell' for the functional basic unit of life. Hooke was among the first who recognized that the gravitational force obeys an inverse square law. He was a talented architect too, and designed the church at willen in Buckinghamshire.

HUYGENS' PRINCIPLE (1678)

"At any point on a wave front may be considered as the source of secondary wavelets that spread out in all directions with a speed equal to the speed of propagation of the waves".

(Wave front is a surface over which an optical wave has a constant phase).

This principle is also used to derive the laws of reflection and refraction.

Huygens's principle can be applied to sound waves and water.

HUYGENS-FRESNEL PRINCIPLE (DIAGRAM 1)

"Every unobstructed point on a wave front acts as a source of outgoing secondary spherical waves".

It is a modified form of Fresnel law which is a method of analysing a wave propagation pattern of light, especially in diffraction and refraction.

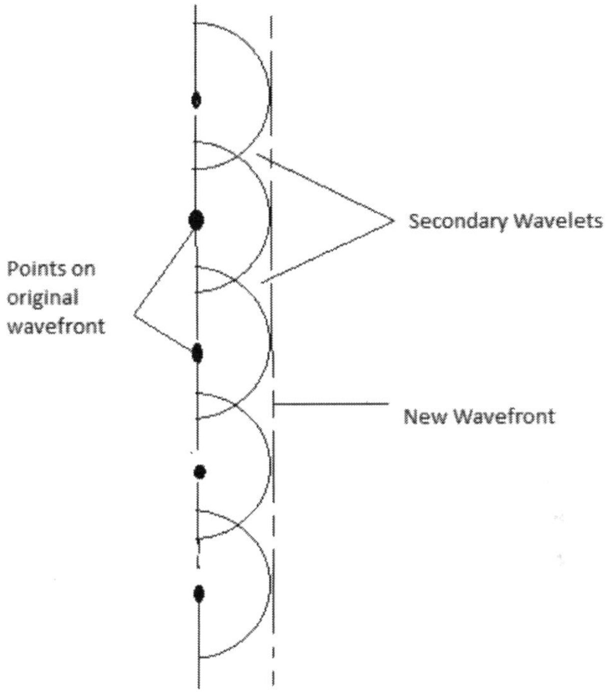

Diagram 1. Huygens's principle for a plane wave.

Diffraction is the bending of a wave around the edge of an opening or an obstacle. This principle in explained by Huygens' Principle and is used to define reflection, refraction and interference. Anything experienced with this phenomenon is called a wave. After extension the diffraction principle becomes very useful for the hologram on credit cards.

HAWKING PROCESS

"Emission of the particles by a black hole is as a result of quantum mechanical effects".

HENRY'S LAW (1801)

"At a constant temperature, the amount of a given gas dissolved in a liquid at equilibrium is proportional to the partial pressure of the gas".

It is a special case of the partition law which applies only to gasses that do not react with the solvent.

IDEAL GAS LAW

"The product of the pressure and the volume of one gram molecule of an ideal gas are equal to the product of the absolute temperature of the gas and the universal gas constant".

Ideal gas is a hypothetical gas which obeys the gas law exactly.

It is a law to calculate Pressure, or Volume, or moles of compound or Temperature.

The ideal gas law is the combination of all the simple gas laws (Boyle's law, Charle's law and Avogadro's law).

$P \times V = n \times R \times T$ – the famous equation, which shows the relationship between measurable properties of an ideal gas.

Here, P is the pressure, V is the volume, n is the number of moles of molecules, T is the absolute temperature and R is the gas constant (8.314 joules per degree kelvin).

[An ideal gas is a hypothetical gas with molecules of negligible size that exert no intermolecular forces. It obeys the gas law exactly…..].

JOULE-THOMSON EFFECT (1852)

"The change in temperature occurs when a gas expands through a valve or a porous plug into a region of lower pressure".

This effect is also known as Joule-Thomson expansion and joule-kelvin effect.

It is a thermodynamic process that occurs when a fluid expands from high pressure to low pressure at constant enthalpy. It is also employed for liquefaction of gases; as it is less expensive, largely used in industries for liquefying. Hydrogen and helium cannot be liquefied by this method.

Example: Refrigeration.

[Enthalpy is a measure of the total energy of a thermodynamic system and is equal to the internal energy of the system plus the product of the pressure and the volume....].

JOULE'S LAWS

Joule's laws are two laws about heat production by an electric current and the energy dependence of as gas to pressure, volume and temperature.

1. The relationship between the heat, generated by an electric current flowing through a conductor is equal to the product of the resistance of the conductor, square of the current and the time for which it flows. The first law is known as Joule effect and also, sometimes called as Joule-Lenz law

2. The rate of production of heat by a constant direct current is directly proportional to the resistance of the circuit and to the square of the current.

Second law states that the internal energy of an ideal gas which is independent of its volume and pressure; and depends on its temperature.

[The internal energy of a given mass of an ideal gas is wholly a function of its temperature......].

JOULE'S EFFECT (1840)

"The heat produced by the passage of a current through an electric current is due to its resistance".

Joule's first law is also known as Joule's effect or Joule's heating. It is a physical law which expresses the relationship between the heats generated by the current flowing through a conductor.

LENZ'S LAW (1835)

"The direction of an induced electric current is always flows in such a direction that it opposes the change in the circuit or the magnetic field that produces it".

This law helps to understand – how the electromagnetic circuits obey the Newton's third law and the conservation of energy.

Applications

In electrical generator, when a generator induces an electric current, the direction of the current opposes the rotation of the generator; therefore, the more electrical energy a generator delivers, the more mechanical energy is required to turn it.

For *e.g.* electromagnetic braking system, induction cooker and back e.m.f. in electric motor.

Stationary electromagnets induce eddy currents into spinning rotor. These eddy currents create magnetic fields which oppose the original change in flux (Lenz's law). This slows the rotor. More force can be applied by increasing the current in the electromagnets.

[Eddy currents also known as Foucault currents, which are electric currents induced within conductors by a changing magnetic field in the conductor].

MACH NUMBER

"It represents the ratio of the speed of an object to the speed of sound in the surrounding medium".

Mach number = velocity of interest / velocity of sound

In 1838 – 1916, Ernst Mach devised the system of speed measurement using the Mach number.

Example:

- Air speed is defined as a Mach number, because the speed of sound differs at different heights and is also it affected by the atmospheric temperature of the air.

- An aircraft flying supersonically (762 m.p.h). at sea level would schedule a higher speed at altitude, the Mach number remains constant to all variables; therefore Mach numbers give an accurate measurement.

- An aircraft is moving twice as fast as the speed of sound is said to be travelling at Mach2.

MACH PRINCIPLE (1870)

"The inertia of a particle is related to the interaction between the piece of matter and the universe".

For *e.g.*

A body in isolation would have no inertia.

MURPHY'S LAW (1942)

"If anything can go wrong, it will go wrong".

This is a supposed law of nature, which is expressed in the popular sayings. This law is also known as an epigram.

In science, this law is complex; but in the common form, it is fairly accurate and sufficient for most applications. There are a number of basic logical reasons for this: *e.g.* nothing lasts forever, so in the long run every part of every machine will break down finally.

MAXWELL'S EQUATIONS (1864)

"These are the set of differential equations which describe the space and time dependence of the electromagnetic field and forming the basis of classical electrodynamics".

The mathematical relationship shows us how to electrify our world and transmit energy and information through the air.

At the beginning of the 19th century, we lit our homes and offices with candles and oil lamps. Communications transformed into handwritten letters which took several days, weeks and months across the country and across the oceans. Today we use electricity in homes and everywhere. We communicate instantaneously to each other, everywhere and anywhere with the radio waves. This enormous discovery was introduced in the 1830s by Michael Faraday who built electric motors and showed that the two natural forces, electricity and magnetism, are related to each other. He proposed that these forces existed as "fields" that permeated space.

Maxwell defined the equations that described these fields. These equations became very popular in understanding the antennas and electromagnetics. He was first to determine the speed of propagation of electromagnetic wave, which is same as the speed of light:

- EM waves and visible light are the same.
- Shorter wavelengths of light include UV and Gamma rays.
- Longer wavelengths include heat (infra-red waves..), microwaves and radio waves.

In 20th century communication methods have been transformed from radio to television and radar to mobile phones are based on electromagnetic fields which were described by Maxwell's equations. These equations are also known as differential and integral forms of Gauss' law of electricity, Gauss' law of magnetism, Faraday's law of induction and Ampere's law.

NEWTON'S LAW OF UNIVERSAL GRAVITATION

"In the universe, any two objects attract each other with a force is directly proportional to the product of their masses and inversely proportional to the square of the distance between them".

This law is about the universality of gravitation. All objects attract each other with a force of gravitation attraction. Gravitation is universal. The force of gravitational attraction is directly dependent upon the masses of both objects and inversely proportional to the square of the distance that separates their centers.

For *e.g.*

- Gravitational fields and tides.

  As we know, the earth experiences two high tides per day because of the difference in moon's gravitational field at the earth's surface at its centre.

  There is a high tide at the moon's nearest side- because the moon pulls the water away from the earth on the far side.

  The tidal influence on a closer object is greater, because the inverse square law drop into a gravitational force which gives a greater ratio of the force on the near side of the object to that on the far side.

  Sun is 391 times as far away from the earth as the moon and its force on earth is about 175 times larger, but its tidal effect is smaller than that of the moon- because tides are caused by difference in gravity field across the earth.

- Newton's law of motion

  "In uniform state, a body tends to remain in that state of motion unless an external force is applied to it".

  For *e.g.*

  In emergency, when we apply the brakes, seatbelt prevents us from hitting the windshield.

- Newton's second law of motion

  "An object's acceleration depends upon the net force acting upon the object and inversely upon the mass of the object".

For *e.g.*

When we push a large and small box with the same force, the small box goes faster.

- Newton's third law of motion

"For every action there is an equal and opposite reaction (Fig. **3**)".

For *e.g.* When we push on a door, it pushes back with the equal and opposite force.

Fig. (3). Rowing boat – Newton's third law (Google image).

OCCAM'S RAZOR

"Entities should not multiply needlessly".

It is a logical principle which states that one should not make more assumptions than required. This principle is also called the principle of parsimony.

If two theories predict phenomena to the same accuracy, then the simpler one is the better.

There are two parts that are considered as the basis of Occam's razor:

- The Principle of Plurality - Plurality should not be placed without necessity.
- The Principle of Parsimony - It is pointless to do with more that can be done with less.

Occam's razor is based on the notion that simplicity is equal to perfection.

OHM'S LAW

"The ratio of the potential difference between the ends of a conductor to the current flowing through it is constant".

$V = IR$

where V is the potential difference between one end to other end of the conductor, I is the current flowing through the conductor and R is the resistance of the conductor.

This is one of the powerful formulae for the studies of electric circuit.

The law gives a simple method for calculating the voltage, current or resistance in a conductor, when two of these are to be known.

Ohm's law is valid for metallic conductors (*i.e.* copper and tungsten) in direct-current circuits as long as the current is relatively low. High currents will heat a metallic conductor and causes its resistance to change; so that the ratio of voltage to the current in the conductor will also change.

For *e.g.* light bulbs, electric stoves, just about any other electrical circuit.

PASCAL'S PRINCIPLE (1647)

"In a confined fluid, extremely applied pressure is transmitted uniformly in all direction". The force acts at the right angle to any surface in contact with the fluid.

For *e.g.* Pump (Fig. **4**).

Pump

Fig. (4). Pump - Applications of Pascal's Principle (law).

PETER PRINCIPLE

"In an organizational hierarchy, every employee tends to rise to his level of incompetence".

This principle is a concept for management theory in which the selection of a candidate for a position is based on their performance in their current role rather than on their abilities relevant to a planned role.

This principle suggests the people that they will be tend to promoted until they reach their “position of incompetence”. This principle is a special case of a present observation – anything that works will be used in progressively more challenging applications until it fails.

We observe it when we applied it to hardware such as vacuum cleaners as aspirators, and administrative devices such as the Safety Evaluations are used for managing change.

PLANCK’S LAW (1900)

“The energy as a quantity radiated by a body in small discrete packets; each of which has energy equal to the product of the Planck’s constant and the frequency of the radiation”. Small packet is known as quanta.

This law describes the electromagnetic radiation emitted by a black body in thermal equilibrium at a definite temperature. It is a fundamental law of quantum theory also known as Planck’s quantum principle.

Planck’s Constant

Energy can be absorbed or radiated by a body in discrete quantities and the energy of a particular radiation is proportional to its frequency. The proportionality constant is called Planck’s constant and is represented by 'h' and its value is 6.626176×10^{-34} joule-second.

It is a physical constant which is used to describe the size of quanta. It plays a central role in the theory of quantum mechanics. This constant is about how the universe is numerically fine-tuned to allow life to exist.

LAW OF REFLECTION

“If a ray of light reflected from a smooth surface, the angle of reflection is equal to the angle of incidence and the incident ray, the reflected ray and the normal ray all lie in the same plane”.

There are two laws of reflection:

1. The angle of incidence is always equal to the angle of reflection.
2. The incident ray, the reflected ray and the normal ray always lies on the same plane.

For a wave-front intersecting a reflected surface, the angle of incidence is equal to the angle of reflection.

REFRACTION LAW (SNELL'S LAW)

Snell's law is an equation that determines the angle at which a ray or beam of light is refracted.

When light enters in a transparent material (*i.e.* glass) at an angle, the direction of light is bent or refracted at a different angle. The angle is determined by the initial angle and the index of refraction of the two materials.

If a straight stick is placed half-submerged in water, the stick appears to bent from the point when it entered in the water (Fig. **5**). This optical effect is due to refraction.

Fig. (5). Refraction of light- "broken pencil" effect.

As light passes through one transparent medium to another, it changes speed and bends. The bending of any object depends upon the refractive index of the mediums and the angle between the light ray and the line perpendicular (normal) to the surface separating the two mediums.

The law of refraction relates the angle of incidence to the angle of refraction.

- Angle of incidence- angle between the incident ray and the normal ray.
- Angle of refraction – angle between the refracted ray and the normal ray.
- Critical angle for refraction -When a light passes from a material of high index of refraction to low index of refraction, there is an angle at which the light starts to be reflected at the interference of the materials. This is called critical angle for refraction.

RELATIVITY PRINCIPLE

Principle of relativity shows the relationship between observations made at different places.

"The space and the time are the relative concepts rather than the absolute concepts".

There are two postulates of relativity proposed by Einstein:

1. The speed of light in a vacuum is constant and independent of the source or observer.
2. All the laws of physics must assume the same mathematical form in any inertial frame of reference; so it is possible to determine the absolute motion of a system by any means.

The extension of this theory which include gravitation and related acceleration phenomena is called general relativity or general theory of relativity.

STEFAN-BOLTZMANN LAW

"Total energy radiated per unit surface area of a black body is directly proportional to the fourth power of its absolute temperature".

This law is also known as Stefan's law. Stefan-Boltzmann constant (related to blackbody radiation) is very useful to measure the amount of heat emitted by a blackbody.

This Law can be applied to a star's size in relation to its temperature and luminosity. It can also be applied to any object emitting a thermal spectrum (*e.g.* metal burners on electric stoves and filaments in light bulbs).

SUPERPOSITION PRINCIPLE

"When a number of influences are acting on a system, the total influence on that system is the sum of the individual influences; that is, influences controlled by the superposition principle add linearly".

The principle of superposition always means the principle of *linear superposition.* The concept is meant for two effects which act on the same thing at the same time; together they give a result that is the sum of the two results which would happen if the two effects acted at different times.

It is also known as superposition property. The superposition principle applies to any linear system including algebraic equations, linear differential equations and system of equations.

The concept means that for two effects which act on the same thing and at the same time, together they give a result that is the sum of the two results which would happen if the two effects acted at different times.

THERMODYNAMIC LAWS

There are four laws to thermodynamics which describe the fundamental physical quantities (temperature, energy and entropy) which identify the thermodynamic systems. These laws also describe that how these quantities perform under various circumstances, and refuse to allow certain phenomena (such as perpetual motion).

1. Zeroth law of thermodynamics

"If two bodies are in equilibrium with a third body, then all three bodies are in thermal equilibrium with each other".

2. First law of thermodynamics

"The change in the internal energy of a system is the sum of the heat transferred to or from the system and the work done on or by the system".

3. Second law of thermodynamics

"The entropy of any isolated system never decreases. In any closed system, the entropy will either remain constant or increases" (Fig. **6**).

Fig. (6). Refrigerator- Second Law of Thermodynamics.

4. Third law of thermodynamics

"The entropy of a system approaches a constant value as the temperature approaches zero".

The first law is the application of the conservation of energy principle to heat and thermodynamic process. This law states that when heat is added to a system, some of the energy remains in the system and some leaves the system. The energy that leaves does work on the area around it; energy that remains in the system creates an increase in the internal energy of the system.

For *e.g.*

If we take a pot of water at room temperature and add some heat to the system; we see first, the temperature and energy of the water increases and second, the system releases some energy and it works on the environment (may be heating the air around the water, making the air to rise).

The entropy is a measure of the disorder of a system containing energy or information. The less ordered the system, the greater will be its entropy. Entropy is always confused with energy. It is very useful in communications and other fields.

The first part of the law says no reaction is 100% efficient. Some amount of energy in a reaction is always lost to heat and also, a system cannot convert all of its energy to working energy.

The second part of the law says a cold body cannot heat up a warm body. Heat naturally flows from warmer to cooler areas; therefore, heat always flows and spreads out to areas with less heat. If heat flows from cooler to warmer areas, it means that it goes against the nature and the system should be corrected in the said condition.

Third law of thermodynamics gives a state known as "absolute zero". This is the bottom point on the kelvin temperature scale. The kelvin scale is absolute, meaning 0° kelvins which is mathematically the lowest possible temperature in the universe. This is about 273.15° Celsius or -459.7° Fahrenheit.

We can see the third law of thermodynamics by thinking about water. Water in gas form has molecules that can move freely. Water vapour has very high entropy (randomness). When the gas cools, it becomes liquid. The liquid water molecules can still move around, but not as freely as gas because they have lost some entropy. When the water cools, it becomes solid ice. The solid water molecules can no longer move freely, but can only vibrate within the ice crystals; the entropy is now very low. As the water is cooled more, closer to the absolute zero, the vibration of the molecule diminishes. If the solid water reaches absolute zero, all molecular motions would stop completely. At this point, the water would have no entropy at all.

[Unit of energy (erg): 6.2415×10^{11} eV (in eV unit) and 1 cm^2 g s^{-2}(in cgs unit)]

UNCERTAINTY PRINCIPLE (1927)

"Momentum and position of a particle cannot both be precisely determined at the same time".

This is a principle related to quantum mechanics given by Werner Heisenberg.

This principle is one of the most famous ideas in physics. It says that there is dimness in nature, a fundamental limit to what we can know about the behaviour of quantum particles and the smallest scales of nature. The principle arises from the wave properties which are naturally exist in the quantum mechanical description of nature.

This principle is an extension of how we see and measure things in the everyday world and is applied in many things that we observe but cannot explain using classical (non-quantum) physics. The uncertainty principle says about vacuums and vacuums are often defined as the absence of everything (not in quantum theory).

VANDER WAALS' FORCE

"It is the sum of attractive or repulsive forces between all atoms and molecules".

These forces are the result of electrostatic attraction between temporary dipoles and induced dipoles, which is caused by movements of electrons in atoms and molecules.

Water molecules in liquid water are attracted to each other by electrostatic forces, and these forces have been described as Vander Waals' forces or Vander Waals bonds.

This force is commonly caused by the attraction between electron-rich region of one molecule and electron-poor regions of another (the attraction between the molecules seen as electric poles). The attraction is much weaker than a chemical bond. These forces are the intermolecular force that cause molecules to stick together in liquid and solid states of matter, and are responsible for surface tension and capillary action. The strength of Vander Walls' force is related to the size of atoms and molecules (the bigger the atom or molecule, the bigger the force will be).

There are three facts, which are the cause of such forces:

- Dipole-dipole interaction, *i.e.* the force between two permanent dipoles (Keesom force).
- Dipole-induced dipole interactions, *i.e.* the force between a permanent dipole and a corresponding induced dipole (Debye force).
- Force between two instantaneously induced dipoles (London dispersion force).

WAVE-PARTICLE DUALITY

"Matter and light exhibit the behaviour of both waves and particles depending upon the circumstances of the experiment".

This principle is the principle of quantum mechanics which says that the light sometimes act like a wave, and sometime act like a particle, depends on the experiment performing. Low frequency electromagnetic radiation tends to act more like a wave than a particle and high frequency electromagnetic radiation tends to act more like a particle than a wave.

Wave-particle duality refers to the fundamental property of matter where, at one moment it appears like a wave and at the other moment it acts like a particle.

In the 1600s, Huygens proposed a wave theory of light while Newton's proposed a particle (corpuscular..) theory of light and after that this theory became very popular and leads.

For *e.g.* ripples in a pond.

Applications

In electron microscope, where the small wavelengths are associated with the electron, can be used to view objects much smaller than what is visible using visible light. It is a type of microscope that uses a beam of electrons to create an image of the specimen.

Neutron diffraction – neutrons diffraction is the application of neutron scattering to determination of the atomic and magnetic structure of a material. Neutron scattering is used in many different scientific fields (food science, drug synthesis, health care and so on…).

WIEDEMANN-FRANZ LAW (1853)

"The ratio of the thermal conductivity of a metal to the electrical conductivity of that metal is proportional to its time".

The relationship is based upon the fact that the heat and electrical transport, both involves the free electrons in the metal. The thermal conductivity increases with the average particle velocity; so it increases the forward transport of energy, but decreases the electrical conductivity when particle velocity increases because of the collision diverts the electrons from forward transport of charge.

[Thermal Conductivity is the property of a material to conduct heat. It is denoted by k, λ, or κ…].

YOUNG'S EXPERIMENT (1801)

This double slit experiment shows the wave nature of light. It is also known as double path experiment in which a wave splits into two separate waves which later combine back into a single wave. The changes in the path lengths of both waves results in a phase shift with creating an interference pattern.

ZEEMAN EFFECT (1896)

"This is the effect of splitting of a spectral line in a spectrum, when the light source is placed in a static magnetic field".

ZELLINGER'S PRINCIPLE (1999)

"Any elementary system carries one bit of information".

CHAPTER 3

Hypothesis

Abstract: A hypothesis is not a prediction or guess. Testing is very important to prove that a hypothetical statement is hypothetical. Sometimes a hypothesis developed a new knowledge. Science proceeds by observation of nature. If hypothesis does not give any observational tests, nothing can be done by scientists. Only argument cannot make a law, theory and theorem. We can consider a hypothesis as an educated guess which must be testable and falsifiable. The scientific method has four steps: observations and description of a phenomenon, formulation of a hypothesis to explain the phenomenon, use of hypothesis to predict the existence of other phenomenon and several experimental tests of predictions. Experiment is only source to prove the truth. A hypothesis alone can teach us something new and alone can give us certainty; these cannot be questioned. This chapter shows the result of investigations and explanations for a phenomenon and a group of phenomenon.

Keywords: Attraction, certainty, data, empirical relationship, ether, experiment, explanation, generalize, guess, hypothesis, mathematics, observation, pilots, prediction, relativity, rules, space and time, subatomic, things, universe.

Using the physical or mathematical rules, a law can be proved but a theory cannot prove.

A scientific law explain things but cannot describe them.

The theory of relativity cannot be proved yet, but alternate theory exists.

A good theory explains all the known data and predicts an unobserved data.

A scientific theory gives a statement about a hypothesis, which is supported with the repeated testings.

THEORY OF RELATIVITY

An empirical relationship (relationship between two or more observed sets of data) differs both from a law and a theory.

Mita Thakur

PARTICLE PHYSICS

We know that the elementary particles of matter are of positively charged protons, neutral neutrons and negatively charged electrons. All matters in the universe are known to be stable, unstable and combination of these three particles, forming a large particle called atom.

At this time, there are four major theories: Newtonian Physics, Relativity, Quantum Mechanics and String Theory.

NEWTONIAN PHYSICS

Newtonian physics is about the energy and mass. With the help of this, it could be easy to describe the orbit of the planets in details. It is about the ordinary world and the ordinary universe.

(Newtonian physics helps pilots to understand how fast they need to travel to take off and to keep flying in the air….).

RELATIVITY

By the end of the 19th century Michelson - Morley experiment tried to find the aether (Ether), but failed. Aether theory proposed the existence of a medium, a space filling substance or field as a transmission medium for the propagation of electromagnetic or gravitational forces.

Albert Einstein explained it with the Theory of Relativity. He showed that light had a fixed speed, that time could shrink, and that matter can affect the space-time universe.

(Objects approached with the speed of light, how light is bent by large masses, like the sun, and how this would affect the nature of space and time…..).

This is the second important theory of physics.

QUANTUM MECHANICS

It describes the behaviour of very small things like atoms and subatomic particles and is used to describe matter on the scale of molecules, atoms or elementary particles. At low temperatures, some phenomenon shows quantum behaviour on macroscopic scale.

There are two types of phenomena at macroscopic scale: waves and particles.

Waves are de-localised phenomena (they are spread out in space) which carry energy but no mass as they move.

For *e.g.*

Ripples on a lake are waves (there is no net transport of water, so no net transport of mass)

Particles are localised phenomena which transport both mass and energy as they move.

For *e.g.*

Cricket balls (Physical objects that one can touch)

In quantum mechanics, in a certain situations particles (*e.g.* electrons) can behave like waves, and waves (*e.g.* electromagnetic radiation: light) can behave like particles. Thus electron can create wave like diffraction patterns upon passing through narrow slits, like water waves, as they pass through the entrance to a harbour and the opposite, photoelectric effect (absorption of light by electrons in solids) can only be explained if the light has a particulate (a substance consisting of separate particles) nature (concept of photons). Such ideas are known as "Principle of Wave-Particle Duality".

STRING THEORY

It is a theory in which particles are described as waves on string. String has length but have no other dimensions.

This theory includes the findings of General Relativity and Quantum Mechanics - how atoms can impact cosmology and how General Relativity can impact the structure of atoms.

These four theories reflect on the variation is in the Universe.

By integrating all into one theory, scientists give a theory, which is called "A theory of everything".

CHAPTER 4

Physics: What We Do and We Don't Know

Abstract: This chapter pays little attention on how physics has given us a whole new point of view; physics behind a series of amazing facts. The definition of physics is not only associated with study of matter, energy and the interaction between them; but it is actually about to asking fundamental questions and try to answer them by observing and experimenting. This chapter includes few topics like: important ideas, legworks, important laws to remember, use of physics, role of physics and a theory of everything. This chapter also gives a look at what a world would be without physics.

Keywords: Amazing facts, atomic nucleus, conservation, cosmology, distance, dynamo, elementary particles, everyday life, fluctuation, general ideas, GRT, ideas, interaction, legworks, matter energy, microscopic scale, QFT, string theory, vibrational mode.

Cosmology and elementary particle physics have largest to smallest ranges.

According to cosmology, the farthest distance light could have travelled since the universe became transparent to light, over ten billion ago; while the elementary particle explores the distances smaller than the atomic nucleus.

The string theory is a theoretical work which says, in particle physics, the point like particles are replaced by on dimensional object, called string. All the different fundamental particles are the example of one basic object that is called string. This theory is a mathematical theory, which tries to explain the standard model of quantum physics; a basic thing (not composed by anything) naturally exists and other things are made from it (like musical string, it can vibrate and each mode can be viewed as a point like elementary particle).

The first string theory was focused only on bosons. The superstring theory integrates fermions and supersymmetry.

The M-theory is a superstring theory, proposed in 1995.

Mita Thakur

String theory is a bridge between two theories GRT (general relativity theory) and QFT (quantum field theory). There are many important ideas, that we know very well and there are many interesting things which we do not know.

IMPORTANT IDEAS IN PHYSICS

This topic gives ideas that are important to describe the physical world around us. Here is a brief of what I believe to be the most important general ideas in physics. For many of those ideas, it took centuries to realize how important they are.

- Atoms and particles:

  Matter is composed of small particles in constant motion, maintaining its stability and do not change places abruptly.

The grains of sand, water or air they all appear in perfect motion.

Atoms consist of electron surrounding a nucleus which contains protons and neutrons.

Particle is an extremely small unit of matter.

- Everything is moving :

  On the microscopic scale, things are never at rest, they are continuously moving around all the time either due to thermal or quantum fluctuations.

- Conservation laws:

  Energy, charge, momentum *etc.* are conserved under appropriate circumstances. It is hard to be accustomed to energy conservation from observing everyday life. As we say one can at least see that energy is neither created nor destroyed.

For *e.g.*

Car petrol tanks needed refill - energy is used up or lost (here fuel is used up). This shows the law of energy conservation.

Electrical energy converts into magnetic energy in the electromagnets.

Mechanical energy converts into electrical energy in dynamo.

These are based on conservation of energy, which says energy can be transformed from one form to another.

- Everything is moving:

  On microscopic scale, things are moving around all the time, they are never at rest. This concept dispute to everyday experiences, where things ted to stop moving due to friction.

- Oscillations and resonances:

  When things are arranged in a stable configuration, they have the tendency to oscillate around that configuration at some specific frequencies. Everyday life gives examples like a pendulum clock, watches (Fig. 7) or musical instruments.

Watch

Fig. (7). Quartz Watch - Physics of oscillation.

- Wave fields and interference:

  Oscillations can propagate through space and they are called waves.

For *e.g.* water waves, matter waves, electromagnetic waves, sound waves are described by similar mathematical equations. Whenever the waves overlap, it shows the important phenomenon of interference.

- Periodical structures:

  Atoms arranged into crystals (periodic in space), oscillations are periodic in time; wave fields can have plane waves that are periodic both in space and time. Periodical structures exist in the macroscopic world (sunflower, caterpillar, *etc.*) but which we observe around us is not periodic.

- Time evolution and causality:

  The present state of the world determines what happens next.

Action equals reaction - however, since the speed of light is so large, in everyday life there seems to be instantaneous action at a distance (*e.g.* magnetic forces or gravitational forces…). As a side-effect of forces being transmitted *via* wave field, the "action equals reaction" first of all applies to the interaction between the particles and the fields.

LEGWORKS

In the recent years there have been a number of achievements in technology. The technological revolutions brought great benefits with environmental and other related problems. New researches and new findings have boosted the technology. Here are some active physical works which are involved in doing something.

- The 'miracle material' graphene

  Graphene is the miracle material of the 21st century. It is a pure carbon in the form of a very thin, nearly transparent sheet. Taken from graphite, it is a good thermal and electrical conductor and can be used to develop semiconductor circuits and computer parts.

- It is faster and cheaper device which is thinner and flexible. It can be used for anything from composite materials like carbon-fibre, used currently - to electronics.

- Graphene can be compared to the way plastic is being used today.

A graphene credit card could store as much information like computers and in future graphene will lead in gadgets like iPhone, kindle, *etc.*

Graphene was discovered in Britain a few years ago. A sheet of graphene is as thin as Clingfilm that could hold the weight of an elephant (assumed).

- 3D Printing: magic process

It is a computer controlled printer; a method of manufacturing everything from shoes to jewellery and from gun to aerospace.

Starting with the bottom layer, wait for dry or solidify and then work with its way on.

The layer process differs on the printer and the material it works with (*i.e.* metal, plaster, polymer, resin). It also depends upon whether it is an industrial or commercial purpose. This type of printer is known as desktop fabrication - it can form any material that can be obtained as powder.

For creating an object, it will need a digital 3D model; which can scan a set of 3D images or draw it using computer assisted design (CAD).

Advantages

– Lower Cost

– Shorter Response Times

– Durability

– Allows businesses to construct working models in just hours

– Superior Surface Finish

– Functional Models

– Teleportation

Teleportation is the theoretical transportation of a matter or energy from one point to another point without traversing the physical space between them. This is a way for superconducting quantum chips to talk to each other over a large distances through an optical fibre allowing quantum teleportation; which is a global quantum internet *via* a quantum repeater. The idea makes use of the tiny magnetic fields generated by these quantum chips to alter the properties of an optical cavity, *via* a magnetostrictive (A material, that is physically expanded in the presence of a magnetic field) material. Superconducting quantum chips are one of the most promising areas of development to become the hardware for tomorrow's quantum computers.

Applications

Quantum computing- quantum teleportation and quantum sensing, would benefit highly from the ability to connect up quantum devices over long distance physics can be seen as the most fundamental of the natural sciences.

In physics and cosmology, we see what happened to our universe as early as a tiny fraction of a second after the big bang that is very surprising to us. Physics divided into several specific fields like electronics, quantum physics, astronomy and biophysics, and so on.

Chemistry is a complex application of physics. It concentrates on the interaction of energy and matter in chemical systems. In chemistry we try to understand the complicated reactions among atoms and molecules. Biology is an application of chemical properties in the living things, which is ruled by the physical laws. In biology, we study how the living cell works and how they designed genome.

Important Laws of Physics to Remember

The laws of physics are considered as the fundamental, but many of them are introduced to be idealized, closed systems and in different circumstances, which are hard to obtain in the real world. Under this topic we discuss the important laws which are must to know about physics.

Laws

- Newton's laws of motion

Newton's three laws of motion describes the basic rules about how the motion of physical objects changes. It defines the fundamental relationship between the acceleration of an object and the total force acting upon it. Sir Isaac Newton changed the understanding of the universe by giving these three laws.

- Newton's law of gravity
- Conservation laws
- Laws of Thermodynamics
- Quantum physics
- Fundamentals of particle physics
- Einstein's theory of relativity
- Newton's law of gravity - which explains the attractive force between the pair of masses. In 12th century, this law was not cleared, because, as Einstein's theory of relativity had provided a more comprehensive explanation for the phenomenon of gravity.
- Conservation laws:

This law states that a quantity or property remains constant during and after an interaction or process.

- Conservation of mass-energy

The total energy in a closed or isolated system is constant.

$E=mc^2$, conservation of mass-energy given by Einstein

For *e.g.*

A nuclear explosion - mass transforms into energy.

- Conservation of momentum

The total momentum stays the same in explosion and collision. It is a fundamental concept of physics along with the conservation of energy and the conservation of mass

For *e.g.*

Safety features in vehicles, seat belt, Airbag

In a car crash, the passenger, the car and its contents experience a great force because of the change of momentum; they slow down rapidly and get injured. If the time taken for the change in momentum on the body is increased, the forces on the body would reduce. Seat belts and crush space are designed to reduce the forces on the body in collision.

In a collision, Seat belt stops the tumbling around inside the car. This increases the time taken for the body's momentum to reach zero, therefore it reduces the force on it. Seat belts are also designed to stretch a bit.

Airbags increases the time taken for the head's momentum to reach zero, therefore it reduces the force on it. It also acts as a soft cushion and prevents injuries.

- Laws of thermodynamics

Thermodynamics is a branch of physics which deals with the energy and work of a system. Energy exists in many forms, such as heat, light, chemical and electrical energy. It only deals with large scale response of a system which we can observe and measure in experiments. The four laws of thermodynamics define the fundamental physical quantities (temperature, energy and entropy) which describe the thermodynamic systems.

1. Zeroth law of thermodynamics

   It states that if the two systems are in equilibrium with a third system, the two original systems are in thermal equilibrium with each other. The law helps to define the concept of temperature.

2. First law of thermodynamics

   It states that energy is always conserved; it cannot be created nor destroyed. Energy can be converted from one form into another form. It is also known as the law of conservation of energy.

   In 1618, Robert Fludd proposed a closed cycle water mill (Fig. **8**) on the basis of first law of thermodynamics.

Fig. (8). Water wheel or Water mill.

Energy created by water passing over a mill wheel exceeds the energy required to get the water back up again with the use of an Archimedes screw.

3. The second law of thermodynamics

   It states that in all energy changes, if no energy enters or leaves the system, the potential energy of the state will always be less than that of initial state. It is called "Entropy" (the quantitative measure of disorder in a system).

   This law describes the limits of what the universe can do. This law is about inefficiency, degeneration and decay. It says all we do is wasteful; there are irreversible processes in the universe.

4. Third law of thermodynamics

   It states that it is impossible to create a perfectly efficient thermodynamic process. This law gives a state known as absolute zero. The third law of thermodynamics relates the entropy (randomness) of matter to its absolute temperature. This law is not famous as other laws. To understand this law we should understand the knowledge of entropy and absolute zero temperature.

- Quantum physics:

It is a branch of physics which deals with properties and behaviour of small particles.

Max Planck first explained the quantum physics and then he developed this with Albert Einstein and Niel Bohr. Quantum physics teaches us that our observations bring the world into existence and it provides us with the opportunity to change both us and the universe. Transistor is the famous of quantum mechanics.

- Einstein's theory of General Relativity

In 1905, Einstein observed that the laws of physics are the same for all non-accelerating observers; so the speed of light in a vacuum is independent of the motion of all observers. This was the famous theory of special relativity, which changed the face of science. This law gives a new concept of space and time. This was given by Einstein in 1915.

There are two main ideas involves in Einstein theory of special relativity.

The principles of relativity - the laws of physics are the same for any inertial reference frame.

1. The principle of the speed of light- The speed of light in a vacuum is the same for all observers, despite of their relative motion or the motion of the source of the light.

Variable speed of light (VSL) is a hypothesis which states that the speed of light (c) may be a function of space and time.

The speed of light in vacuum is considered as a constant (299,792,458 meters per second)

Use of Physics in Daily Life

This topic shows the important science impacts in everyday life. The ideas based on physics plays a huge role in the daily life. Lightning, oils for fuel is extracted and processed through physics based ideas which power up our planes, trains and automobiles, mining coals for factories, in measuring quantities of different solids, the internet, *etc.* Without Physics we cannot live; it is all around us every day and every moment. Here are few example concerned with the physics and its laws.

- When we walk or run, our motion is a part of laws of mechanics and thermodynamics.
- When we eat food, the chemical reaction produces heat energy which converts into mechanical energy.
- When we speak, sound properties like pitch and intensity are produced.
- Cycling a bike is a mechanical movement.
- Boiling water in a kettle, toasting bread is concerned with heat energy and electricity.
- Electricity that we use is a gift of physics.
- Automobiles are based on physics.

- Using the brake is concern with friction.
- The uses of refrigerators, pressure cookers, washing machines, televisions, music systems, computers, *etc*. are designed on the basis of physics.
- Using a door handle is concerned with lever and spring.
- Throwing a ball is explained with weight, gravity, trajectory, air resistance.
- According to Newton's third law, when we walk, we push the ground backwards, and the ground pushes us forward.
- An airplane flies due to application of Bernoulli's theorem, based on conservation of energy.
- We tend to move forward when sudden brakes are applied - Newton's law of inertia
- While riding in a circular path, we tilt towards the centre; it is due to centripetal force.
- When a hot cup of tea is left for cooling, its rate of cooling varies directly with the temp of the surroundings. This is Newton's law of cooling.
- We stay stuck to the earth due to gravitational force of the earth.
- The frequency of sound of a distant vehicle coming towards us or going away from us changes, this is due to Doppler Effect.
- We see this world colourful, because every object reflects light of a typical wavelength, associated with a unique colour.
- We see a rainbow, due to the dispersion of light by water droplets in atmosphere.
- We feel warm when standing in sunlight, because the sun sends energy to us in form of electromagnetic radiations, which gets converted into heat.
- Solar cells.

- Solar cell or photovoltaic cell is an electrical device which converts the energy of light directly into electricity by the photovoltaic effect. It explains the physical process by which light is converted into electrical current when striking a suitable semiconductor device (Fig. **9**).

Fig. (9). Solar power system - photovoltaic (PV) panels converts sun's rays into electricity.

- These are utilized in daily life through several different technologies:
- Active solar technology

  It is a solar collecting device, which requires energy besides the sun to function and can be store or transfer heat energy in water or air

- Passive solar technology

This is a moving part of the system involved with active solar technology.

- Solar thermal

It generates heat by collecting heat energy directly from the sun.

- Photovoltaics (PV)

It is a method for converting sun's energy into electricity, which is known as photovoltaic process.

Silicon based solar cell collect the sun's light to create DC current; then an inverter converts it into standard AC currents.

The solar panel electricity systems are also known as solar photovoltaics, which capture the sun's energy using photovoltaic cells. These cells do not need direct sunlight to work; they can also generate some electricity on a cloudy day. The cells convert the sunlight into electricity which can be successfully used to for household electricity.

[PV cells are made from layers of semiconducting material (usually silicon). When light shines on the cell, it creates an electric field across the layers……].

- Direct and indirect solar energy

Direct solar energy requires one step for converting the sun's energy into heat and electricity.

Today solar energy systems are one of the cleanest powers generating technology. They convert the sun's rays into electricity or hot fluids, without creating air pollution, hazardous waste or noise.

- Big Bang theory

It explains how the universe began; what happened at the very beginning of our universe. There are many misunderstanding regarding the big bang theory. According to this theory About 13.7 billion years ago in the universe all the matte was concentrated into a single tiny point. This began to enlarge rapidly in a hot explosion and still expanding. The theory is incomplete without asking the question - what is God? This is because cosmology (the study of the origin of the universe) is an area where science and theology meet.

Role of Physics in Daily Life

This topic shows an increased understanding of the concepts of physics in the 20th century; the development of products that have transformed the way we live our lives.

Man has always been inspired by the wonders of nature. He has always been eager to know about the secrets of nature and remained in search. In our daily life, we hardly find a device where physics is not involved. Health, economic development, education, energy and the environment; all are cherished with physics. Now the modern world is much more connected than previous times; we travel far, communicate easily and quickly, conduct business around the world effortlessly. There is almost no place on the earth where the modern world is not connected with each other. We are not only connected with physics through modern technology; we are much more connected with basic level through Mother Nature. The tsunami is a best example. Physics has played an important role in the connectivity of all parts of the planet. Throughout the developing world; they are directly bound with disease, poverty and the degradation of the environment. Lack of access to physics, other science and technology are the main issues for these problems.

We know physics is the study of natural phenomena; here we can see how the technology is related to physics, which we use in our daily life:

- Television uses electromagnets to direct electrons on a screen to produce pictures.
- A cell phone uses microwaves.
- A car is built on mechanical principles.
- A refrigerator uses thermodynamic principles.

- A home uses electricity to power supply.

  Physics is also a necessity in solving our future problems.

  Politics, socio economic factors and acceptance by the phone, all play a role in the development of a nation. Physics, engineering and other technological and scientific achievements can transform a developing nation to a developed nation. Physics can play an important role in developing layout to fight climate change in the development of cleaner energies, technologies, *etc*.

- The gold nanoparticles emitting alpha radiation could be the greatest treatment for cancer.

- Black hole

  An object in space collapsed under its own gravitational forces to such an extent that its escape velocity is equal to the speed of light. It is a region of space which has a intense gravitational field where no matter or radiation can escape; was predicted by the theory of general relativity and known as a super massive object. Because no light can get out, people could not see black hole. Scientist think that the smallest black hole was formed when the universe begun.

- White hole

  A region of space-time that is the time reversal of a black hole; *i.e.* matter would explode out of a white hole.

- White noise

  It is a random noise in which the energy per unit bandwidth is constant for the whole frequency range of interest.

- Higgs boson (1929)

  It is a part of theory which was first proposed by Peter Higgs and then explained (how particles obtain mass) with others in 1960. It is a spin zero particles with a non-zero mass. This is an elementary particle explained in 1964.

- Anthropic principle

"Why is the universe itself just right for life?"

Anthropic is relating to human being or their existence; and principle means law. So it is the law of human existence.

It says the observable universe has to be as it is, rather than any other way, otherwise we would not able to observe it. In astrophysics and cosmology, this principle is a philosophical idea.

- Big bang

It is a cosmological theory says that in the universe, matter and energy were originated from a state of huge density and temperature at a finite moment in the past, when space and time came into existence. It is the singularity at the beginning of the universe which explains the origin of the universe.

- Big crunch

Big crunch is the singularity at the end of the universe- says that the expansion of the universe is due to big bang and it will not continue forever. At certain point in time it will stop expanding and collapse into itself pulling everything in itself and turns into a biggest black hole for ever.

- Cosmological constant (1917)

It is a homogenous energy density that causes the expansion of the universe to accelerate. Einstein introduced this into his equations for general relativity.

- Cosmology

It is the study of the nature, origin and evolution of the universe.

- Cosmological red shift

A shift or an effect where light emitted from a distant source appears red shift because of the expansion of space-time itself.

- Dark energy

  Dark energy is a hypothetical form of energy in the universe linked with the expansion of the universe. It spreads through all the space and tends to accelerate the expansion of the universe.

- Singularity

  A point in space, at which, the space and time curvature (or some other physical quantity) becomes infinite. The existence of singularities is assumed as a limitation of general theory of relativity.

- Special theory of relativity (SR)

  An accepted theory of physics based on the relationship between space and time; the idea that the laws of science should be the same for all observers, no matter how they are moving in the absence of gravitational phenomena. It is all about what's relative and what's absolute about time, space and motion.

- Radar

  An object-detection system - a method of detecting the presence, position and direction of motion of distant objects (like ships, aircrafts) by means of their ability to reflect a beam of electromagnetic radiation of centimetric wavelengths. It is also used for navigation and guidance.

A Theory for Everything

This topic shows a basic law of physics. Physics is the systematic study of natural world. It expresses reality through an accurate application of observation with logics and reasons. The aim of the physics is to develop a error free model of physical reality.

At a fundamental level, all the laws of physics can be sum up just in two formulas:

1. Einstein's theory of gravity gives a cosmic description of the very large (galaxies, black holes and big bang, *etc*...).

2. The quantum theory gives a microscopic description of the very small (microcosm of subatomic particles and radiation...).

The two theories use different mathematics and different physical principles to describe the universe in their areas *i.e.* the cosmic and the microscopic.

The superstring theory unites the gravity with a theory of radiation, which is required to solve the problem of quantum wormholes. It explains the impossible quantum laws of subatomic physics by proposing that the subatomic particles are just resonance or vibration of string.

[Vibration of a violin string corresponds to a musical notes is just like the vibrations of a superstring similar to the particles present in nature; the universe is then a symphony of vibrating string. As a string moves in time, it wraps the fabric of space around it and produces black holes, wormholes and other solutions of Einstein's equations].

According to Einstein theory of special relativity, the speed of light can never change - it is always stuck at appx. 300,000,000 meters/second, no matter who is observing it: nothing can move faster than light - loophole for quantum mechanics. Quantum mechanics is the study at a microscopic scale. These types of particles are impossibly small but very important for everything in the universe.

World without Physics

This topic describes how the world would be without physics; Nonexistence. If our space-based functionality does not exist; stock markets, the internet, cashpoint machines and mobile phones would all stop working. What happen if we fall into black hole?

Physics is the study of the physical world around us; a world without physics would be a world without physical form. Without the knowledge of physics, the application of physics (electronics, mechanics, *etc.*) would not exist. We would live without developed technology or advanced tools such as hammer, watch… if one could stop time? If Wind stops blowing, water stops running, and the Earth stops turning? Everything is at a stand-still?

The atmosphere is frozen and we are stuck in position, unable to move around the molecules in the air. We would be unable to breath; there is no way to take the motionless air into lungs, we try to call out for help and find that no sounds can resonate through environment being incapable of using the atmosphere as a transmitter of sound waves. We freeze to death- the world having ceased to emit any heat.

How time works and what it really is. It is a great mind exercise to think about things like traveling, realizing and stopping time.

Without the law of physics, nothing would exist. Our world exists due to Gravity, inertia, energy becoming mater, gasses becoming liquids at high pressure and liquids boiling at high temperatures, *etc.*

The importance of physics includes awareness of the effects of natural events. When tsunamis start to rise and threaten disasters, it is the knowledge of physics that helps to know which areas of the world's population to warn and the proper precautions that they must take. The principles of physics that involve planetary movements operate the design and function of manmade satellites. Without knowledge of the basic principles of motion and force, humankind's existence would still be ancient. Without it we cannot maintain our health, cannot get energy for everyday use, cannot communicate easily, cannot transport easily, and cannot develop electronics, constructions, and economy. It is amazing.

CHAPTER 5

About Physics

Abstract: Physics is not only the laws, theory and theorems; it is about to asking the fundamental questions and answering them by observing and experimenting. This chapter shows what physics is? How it works and why it is necessary to everyday life? Physics is concerned with understanding the world around us. One of the best ways to clarify our thinking about the visionary problems is through physics; to analyse ideas and arguments, to provide reasons and evidences for judgements and to explain. This area includes few topics like physics myths, why physics helps to understand world, how physics is useful to help others and few applications and examples with illustrations.

Keywords: Analyse, application, arguments, arguments, evidences, ideas, ideas, illustration, information, macro fundamental question, misconception, mythology, myths, popular, random, reason, reasons and evidences, truth, visionary problems, world.

When we think **why sky is blue?** Physics gives us the answers:

The sky looks blue due to Rayleigh scattering.

Sun's white light is a mixture of all colours. It was demonstrated by a prism. Prism separates different colours, which form a spectrum. Different colours have different wavelengths. The visible parts of the spectrum are from red to violet known as VIBGYOR (violet, blue, green, yellow, orange, red) depending upon wavelength ranges. The colour receptors in the human eye respond most strongly to the red, green and blue wavelengths and these colours give us our colour vision.

In the air, molecules scatter blue light more than red light from the sun. As light moves through the atmosphere, most of the longer wavelengths pass straight through and little (red, orange and yellow) of them is affected by the air; much of the shorter wavelength light is absorbed by the gas molecules. The absorbed blue light is then radiated in different directions and it gets scattered all around the sky and sky looks blue.

Mita Thakur

The sky looks paler or white closer to the horizon, because the scattered light when passes through more air and then again, it scattered away in other directions. So, less blue light catches the eyes and sky looks paler.

In 1859, the colour of sky was first described and explained by John Tyndall (known as Tyndall effect); after that it was studied by Lord Rayleigh in details and named as Rayleigh scattering. They thought that the blue colour of sky is due to the small dust particles and the droplets of water vapour in the atmosphere [Wavelengths of VIBGYOR colours range from 390 to 700 nm].

WHY SUNFLOWER LOOKS YELLOW?

A yellow sunflower reflects yellow wavelengths while absorbing all the other colours.

Asking, observing and experimenting and answering are physics.

The Mythology of Physics

There are few myths and misconceptions about popular Physics:

- Physics is too hard - avoid it
- Newton's Apple - false story
- The Ice diet - eat ice loose calories
- Quantum physics - the world is completely random.
- The principles of quantum physics apply at the everyday "macro" level of reality.
- Quantum physics proves that the mind directly creates and shapes reality.
- Quantum physics allows for the existence of ghosts and other non-corporeal entities.

- The Larger Hadron Collider will destroy the earth.
- The second law of thermodynamics falsify the evolution (concept of entropy).
- Evolution causes something to go from lower to higher.
- Light is both a wave and a particle at the same time.
- Noise travels in space.
- There is no gravity in space.
- When exposed to the vacuum of space, the human body pops.
- Polaris is the brightest star in the northern hemisphere night sky.
- There is a dark side of the moon.
- Meteors are heated by friction when entering the atmosphere.

Why Physics?

The turning point of this topic is how physics helps us to understand the world around us. Physics is essential to understand this, the world inside us, and the world beyond us. It is the most basic and fundamental science.

From light bulb and cell phone to muscle, lungs and brains, from earthquakes, tsunamis and hurricanes to quarks, DNA and black holes; it help us to organize the universe.

- It deals with fundamentals and helps us to see the connections between satisfying appearances of different phenomena.
- It gives powerful tools to express our creativity; to see the world in new ways.

- It gives quantitative and analytic skills needed for analysing data and solving problems in the science, engineering, medicine, economics, finance, management, law and public policy.

- It is the basis of modern technology, tools and the instruments, which is used in the researches and developments. Manufacturing is managed by physics based technology.

- Colour television

  Colour television is a television transmission technology. In this system, camera filters the light from the screen into three primary colour components, red, blue and green, which are detected by separate camera tubes. This separate information obtained, relating to the colour of the image, is combined with the sound and synchronization signals and then transmitted using the system. At the receiver, the signal is split again into red, blue and green components, each being fed to a separate electron gun in the cathode ray tube of the receiver. By an additive process the picture is reconstituted by the beam from each gun activating a set of phosphor dots of that colour on the screen. It is an improvement on the earliest television technology.

- Computer

  A computer is a general purpose device that processes information according to a set of instructions, called program. Modern computers are based on integrated circuits.

- Hubble telescope

  This is the world's first space based optical telescope. Dr. Edwin P. Hubble established the truth about an expanding universe, which gives the foundation for the Big Bang theory. He found many galaxies in the universe. His work gave the idea that the universe was getting bigger, as if from a big explosion. Scientists call this explosion the big bang. The Hubble Space Telescope is a large telescope in space. It was launched by NASA in 1990. It is the length of a large bus, weighs as much as two adult elephants and travels around Earth at about 5 miles per second. It takes pictures of planets, stars and galaxies.

- Nuclear power plant

  It is a thermal station in which the heat source is a nuclear reactor: heat is used to generate steam which drives a steam turbine connected to a generator, which produces electricity.

- Nuclear weapon

  Nuclear weapon is a device, which derives its destructive force from nuclear reaction. In this reaction, huge amount of energy releases from small amount of matter.

- Nuclear reactor

  It is a device to initiate or control a sustained nuclear chain reaction: nuclear fission chain reaction is sustained and controlled in order to produce nuclear energy, radioisotopes or new nuclides.

Reactors are used for generate electricity, moving aircraft carriers and submarines, for produce medical isotopes for imaging and cancer treatments and for conduct researches.

How Physics Helps Us to Others?

This topic is concerned about 'how physics works for us and how we help others'. It explains us how universe behaves, how the basic objects and forces interact? There are the countless questions. Physics is not only relevant to the big questions of from where the universe came and how the universe works; it has many real life applications few are seen below.

- Physics is all over around us.

  e.g. The electric light, the vehicle, wristwatch, cell phone, CD player, radio, plasma TV, computer, *etc.*

- Physics helps us to understand the mysteries of universe.

- Without physics medicine technology would be in dark.

- Physics is necessary for studying of patterns in the physical interactions of the observable universe. Without physics many applications like electronics, mechanics and so on-would not exists. We would live without developed technology or advanced tools.

  Without the laws of physics, nothing would exist. It is the heart of all, it is totality, it is everything, and it is existence.

Few Applications

This area gives some important applications (Fig. **10**) which is very useful for us. Many applications of physics use concepts from a range of topics covered by physics. Every technology involves physics. While Physics deals with the natural world; technologies try to apply this on manmade world. Now a days the important branches of physics (such as quantum, thermal, electronics, optics, optoelectronics, electricity and magnetism, computer programming, business, industrial process control, signal and image processing, *etc*...) as well as special topics (green energy technology and geographical information system...) have become more creative with the developing ideas of physics.

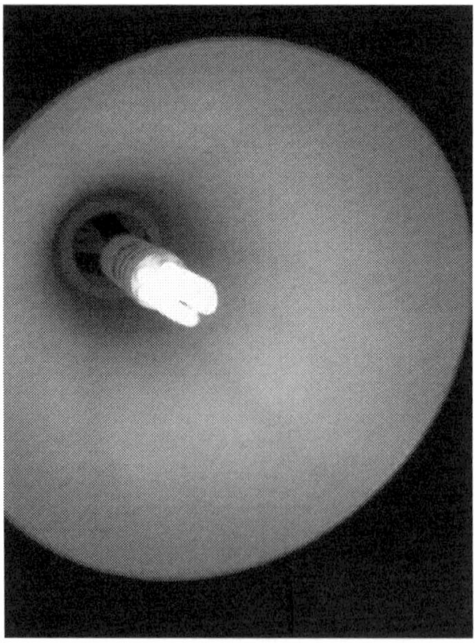

Light Bulb

Fig. (10). Light bulbs convert electricity into light and heat. There are several kinds of light bulbs.

Where physics helps us in various ways, there are some harmful effects which affect us balefully:

- Radio waves - harm body cells, extensiveness of migraine and headache disorders.
- Microwaves - internal heating of body tissues.
- Infrared - skin burns.
- Visible light - increased rates of premature skin aging and skin cancer.
- Ultraviolet - damage to surface cells (including skin cancer) and blindness.
- X-rays - damage to cells.
- Gamma rays - cancer, mutation.

Below are a few everyday examples of physics which are based on applications of physical laws:

- Steaming the food - specific latent heat

  The specific heat of vaporisation for water is large. A plate filled with food absorbs heat from the hot steam. The condensation of steam at the base of the plate releases a large quantity of heat and thus allows food to be steamed.

- Cooling drinks with cold water and ice - specific latent heat

  A glass of hot water can be cooled faster by adding cold water or ice into it. During the melting of ice, a large quantity of specific latent heat is absorbed from the drink and this causes the drink towards a temperature that approaches the melting limit of ice. Ice absorbs a large quantity of latent heat during the process of melting.

- Effect of resonance

  The radio and the television tuner provide the facility to select the programmes. The circuit inside the tuner is adjusted until the resonance is achieved at the frequency transmitted by a particular selected station, and a strong signal is produced. The loudness of music, which is produced by a musical instrument (*e.g.* flute) is also due to resonance in the air.

- Reflection of waves

  The rear and side view mirrors in a car are used to view the cars behind and at the side while overtaking another car; making a left or right turn and parking the car. The mirror reflects light waves from other cars and objects into the driver's eyes for safety.

- Defence

  A periscope is an optical instrument. It is constructed by using two plane mirrors for viewing objects beyond obstacles. The light waves from an object which is incident on a plane mirror are reflected twice before entering the eyes of the observer.

- Telecommunication

  The infrared waves from the remote control of electrical equipment (television or radio) are reflected by objects in the surroundings and received by the equipment.

- Refraction of light

  A swimming pool looks shallower than real - the apparent depth of a swimming pool.

- Rainbow formation

  When sunlight shines on raindrops, refraction and total reflection occur in the raindrop. When an observer receives the refracted light from the rainbow at specific angle, a vision of rainbow is formed.

- Binoculars

  Binoculars are designed to serve the viewer to use both eyes at once when viewing distant objects. There are two prisms are mounted in each half of the binocular. Light rays from distant objects undergo total reflection in the prism before entering the eyes of the observer. The image seen by the observer is erect. Binoculars give users a three dimensional image. These are used in biology and surgery to obtain a stereoscopic view of small objects or parts.

- Drinking with one straw - application of atmospheric pressure

  When we drink the water from straw, we create a vacuum by sucking the air out of the straw. After sucking all the air out of a straw, there is still air pressure surrounding the straw and it is greater than the pressure inside the straw, which then causes the liquid to go up the straw and into the mouth.

- Nuclear weapon

  It is an explosive device, where explosion is caused either by nuclear fission, nuclear fusion or a combination of both fission and fusion.

- A car going down the street - friction between the rubber and the road.
- A leaf falling from a tree - due to gravitational attraction.
- An incandescent light - electrical resistance.
- Florescent light - ionized particles.
- A piece of wood floating in a river - density and buoyancy.
- Drops of water - surface tension.
- A credit card - electromagnetic induction.
- The guitar - waves and sound.
- Ice skating - linear and angular momentum.

- Radar - principle of reflection of electromagnetic waves.

Few Examples

Most common real Life examples of physics:

In essence, physics has made our lives very comfortable. Modern age is the age of science and technology; the technical development is all due to progress of science. The laws of science have been applied in every part of the life. From industries to daily household appliances, in the field of defence, medical, education and energy production; it strengthen the life of people. Life becomes very easy with the involvement of different field of physics. The production of electricity with the help of other energies is due the laws and theories of physics. The solar cells, the turbines, the nuclear reactors and various important projects, all are the contributions of physical laws. The modern industries, machineries all are based upon the basic principle of machines. Physics have played an important role in the inventions of devices in medical science. Radar, laser, jet planes and missiles are only because of the physics. A computer brings the extensive changes in all sectors. Designing, constructions, the field of engineering and medicine, the accountancy and commerce, all are occupied by physics. The kitchen appliances (*i.e.* refrigerator, mixture and blenders), fan motors, microwave ovens, television, radio, air conditioner, telephones, *etc.* for daily uses, all are manufactured by using physical laws. Physics plays a fundamental role in developments.

eBOOK SUMMARY

The phenomena in nature were roughly divided into classes like heat, electricity, mechanics, magnetism, properties of substances, chemical phenomena, light, X-rays, nuclear physics, gravitation, *etc.* The world around us, the world inside us and the world beyond us - physics teaches you to think.

It explains - why the sky is blue? (p. 85)

In the physical world, gravity, force and the other elements of physics rule our abilities and it can be utilized to enhance our every moment. It is crucial to ask people questions and explain things to people- in both cases, the great way is to learn books.

The basic understanding of science is necessary for individuals to take part intelligently in modern social and political conversations. I have seen the questions and thought what book is suitable to read to develop an interest in physics to fill the gap between interests to understanding. A little mind blows in places but will certainly open the eyes on the physics theories and principles such as cosmology, the famous equation $E = mc^2$, *etc.*

There are many different theories which describe the different phenomena of physics.

There are different theories describing the different phenomena of physics. Here I am trying to list them simply without trying to precisely express them. The Laws, Theories and Theorems become very useful to understand the historical perspectives which have led us to our current understanding. It also involves a touch of history, a bit of mathematics, and many everyday examples. In the high technological society it is very important to know the basic understanding of 'how and why the world works'.

The eBook lists the theories and theorems, which describe a basic understanding of how and why the world works, how we benefit from physicist's thoughts and how physics will change the world. This book does not have lots of things and details but does not have the sort of in-depth material which one need to learn and understand.

This eBook is for "general" reading. The target audience is broad and suitable for wonted reading by other scientists and by non-scientist friends. A key target audience is "young science students". It may help readers to understand the underlying principles of Physics for every day.

In short: Keep reading, develop interest and make progress.

PHYSICAL CONSTANTS AND ASTRONOMICAL CONSTANTS

Much of the work in physics has allowed the results to take the form of constants. The certain values are relatively unchanged and can integrate themselves into new problems to solve.

Formal sets of constants, along with recommended values

Table 1. Physical constants.

Quantity	Symbol	SI Unit Value	Alternate Unit Value
Avogadro constant	N, L	6.022169×10^{23}	6.022169×10^{23}
Atomic mass unit	u or Da	1.67377×10^{-27}kilogram	1.67377×10^{-24}gram
Boltzmann's constant	k	1.3807×10^{-23} joule/kelvin	1.3807×10^{-16} erg/kelvin
Electron radius	r_e	2.81794×10^{-15}meter	2.81794×10^{-13}centimeter
Earth's mass	m_{earth}	5.974×10^{24}kilograms	5.974×10^{27}grams
Faraday constant	F	9.65×10^{4}coulombs / mole	--
Gas constant	R	8.3145 joules / kelvin /mole	8.314 x 107 ergs / kelvin/mole
Gravitational constant	G	6.67×10^{-11} Newton ms/kgs	--
Moon's mass	m^{moon}	7.348×10^{22} kilograms	7.348×10^{25} grams
Planck's constant	h	6.626176×10^{-34} joule-sec	6.626176×10^{-27}erg-sec
Neutron radius	r_n	*	*
Proton radius	r_p	2×10^{-16} meter*	2×10^{-14} meter*
Sun's mass	m_{sun}	1.989×10^{30} kilograms	1.989×10^{33}grams
Speed of light	c	2.99792×10^{8}meter /second	2.99792×10^{10}cm/sec
Speed of sound in air	v_{sound}	331.4 m/sec	1087 ft/sec
Wien's constant	σ_w	.898 x 10-3 meter-kelvin	0.2898 cm-kelvin

ASTRONOMICAL CONSTANTS

Table 2. Astronomical constants.

Quantity	Symbol	Value
Gaussian gravitation constant	k	$0.017\ 202\ 098\ 95\ A^{3/2}\ S^{-1/2}\ D^{-1}$
Constant of gravitation	G	$6.673\ 84 \times 10^{-11}\ m^3\ kg^{-1}\ s^{-2}$
Mass of Sun	S	1.9818×10^{30} kg
Hubble constant	H_o	$70.1\ kms^{-1}\ Mpc^{-1}$
Speed of light	c	$299\ 792\ 458\ ms^{-1}$
Jupiter radius	R_J	69,911 km
Solar luminosity	$L_\odot$	3.839×10^{33} erg s-1
Astronomical unit	AU	149 597 871 km
Parsec	pc	$3.08567758 \times 10^{16}$ meters
Light year	ly	9.4605284×10^{15} meters
Year	yr	3.156×10^{7} s

Bibliography

Ames, J. S., & Benton, J. R. (1904). Text-book of general physics. New York: American Book.

Appleby, S., & Linder, E. V. (2012). The paths of gravity in galileon cosmology. Journal of Cosmology and Astroparticle Physics, 2012(03), 043-043. doi: 10.1088/1475-7516/2012/03/043

Berman, B. (n.d.). Zoom: How everything moves: From atoms and galaxies to blizzards and bees.

Bird, M. L. (2002). What's New in FOCUS and at The American Geographical Society? Focus on Geography, 46(4), 1-1. doi: 10.1111/j.1949-8535.2002.tb00019.x.

Boozer, A. D. (2013). Time-reversal invariance and time asymmetry in classical electrodynamics. American Journal of Physics, 81(8), 585. doi: 10.1119/1.4807756

Bor, D. (2010). The Mechanics of Mind Reading. Scientific American Mind, 21(3), 52-57. doi: 10.1038/scientificamericanmind0710-52

Brown, I. A., & Hammami, A. (2012). Gauge issues in extended gravity and () cosmology. Journal of Cosmology and Astroparticle Physics, 2012(04), 002-002. doi: 10.1088/1475-7516/2012/04/002

Cassidy, D. C., Holton, G. J., & Rutherford, F. J. (2002). Understanding physics. New York: Springer.

Cassidy, D. C., Holton, G. J., & Rutherford, F. J. (2002). Understanding physics. New York: Springer.

Coleman, P. (2011). Introduction to many body physics. Cambridge: Cambridge Univ. Press.

Contents: Laser Phys. Lett. 3/2012. (2012, 12). Laser Physics Letters, 9(3), 169-174. doi: 10.1002/lapl.201290004

Cowley, S. C. (2006). Plasma physics: Thermonuclear ringtones. Nature Physics, 2(5), 299-300. doi: 10.1038/nphys308

Das, A., & Ferbel, T. (2003). Introduction to nuclear and particle physics. River Edge, NJ: World Scientific.

Dine, M., Banks, T., & Sachdev, S. (2012). String theory and its applications: From meV to the Planck scale: Tasi 2010, proceedings of the 2010 Theoretical Advanced Study Institute in Elementary Particle Physics, Boulder, Colorado, 1-25 June 2010. Singapore: World Scientific.

Mita Thakur

Dolan, J. F. (1970). A catalogue of discrete celestial x-ray sources. Woodbury, NY: Published for the American Astronomical Society by the American Institute of Physics.

Daintith, John. A Dictionary of Physics. Oxford: Oxford UP, 2009. Print.

Eddington, A. S. (1928). The nature of the physical world. New York: Macmillan.

Editorial. (1999). Physics of the Earth and Planetary Interiors, 111(1-2), 1. doi: 10.1016/S0031-9201(99)00025-4

Electromagnetic waves. (1971). Harmondsworth: Penguin, publ. for Nuffield Advanced Science.

Fedorov, A. (2010). The all-organic route to doping graphene. Physics, 3. doi: 10.1103/Physics.3.46

Feynman, R. P., Leighton, R. B., & Sands, M. L. (1963). The Feynman lectures on physics. Reading, MA: Addison-Wesley Pub.

Frauenfelder, H., & Henley, E. M. (n.d.). Nuclear and particle physics. Reading, MA: Benjamin.

French, S. (2011). Deep-sky wonders: A tour of the universe with Sky & telescope's Sue French. Richmond Hill, Ont.: Firefly Books.

Friberg, S. R. (1993). Demonstration of colliding-soliton all-optical switching. Applied Physics Letters, 63(4), 429. doi: 10.1063/1.110013

Garmire, E. (2013). Nonlinear optics in daily life. Optics Express, 21(25), 30532. doi: 10.1364/OE.21.030532

Giancoli, D. C. (1998). Physics: Principles with applications. Upper Saddle River, NJ: Prentice Hall.

Godin, B. (2002). Outline for a History of Science Measurement. Science, Technology & Human Values, 27(1), 3-27. doi: 10.1177/016224390202700101

Google. (n.d.). Retrieved July 05, 2014, from https://www.google.co.uk/

Griffiths, D. J. (1987). Introduction to elementary particles. New York: Wiley.

Haler, P. J., & Stuart, A. H. (1921). An introduction to physics for technical students. London.

Hameka, H. F. (1981). Quantum mechanics. New York: Wiley.

Hawking, S. (1988). A brief history of time: From the big bang to black holes. Toronto: Bantam Books.

Hawking, S., & Mlodinow, L. (2010). The grand design. New York: Bantam Books.

Hogg, D. W. (1997). Near infrared imaging of the Hubble deep field with the Keck telescope. Washington, D.C.: National Aeronautics and Space Administration.

Hohenemser, K. H. (1968). New Developments in Supersonic Transport. Scientist and Citizen, 10(2), 42-43. doi: 10.1080/21551278.1968.9958596

Infeld, L. (1950). Albert Einstein: His work and its influence on our world. New York: Scribner.

Isaacson, W. (2007). Einstein: His life and universe. New York: Simon & Schuster.

Kaku, M. (2014). The future of the mind. the scientific quest to understand, enhance, and empower the mind. New York: Doubleday, a division of Random House, LLC.

Kandel, E. R. (2006). Essay: The New Science of Mind. Scientific American Mind, 17(2), 62-69. doi: 10.1038/scientificamericanmind0406-62

Karam, L. (2014). Measurement traceability in medical physics. Journal of Medical Physics, 39(1), 1. doi: 10.4103/0971-6203.125470

Kardar, M., & Chandler, D. (2009). Statistical Physics of Particles and Statistical Physics of Fields. Physics Today, 62(5), 55. doi: 10.1063/1.3141944

Khan Academy. (n.d.). Retrieved July 05, 2014, from https://www.khanacademy.org/

Kim, J., & Leal, L. G. (2012). Announcement: New Format for Physics of Fluids. Physics of Fluids, 24(1), 010202. doi: 10.1063/1.3659012

Kompaneet□s□, A. S. (1964). Theoretical physics. New York: Gordon and Breach.

Kuhn, K. F. (1996). Basic physics. New York: J. Wiley.

Lauretta, D. S., & McSween, H. Y. (2006). Meteorites and the early solar system II. Tucson: University of Arizona Press.

Ledingham, K. (2005). Laser Induced Nuclear Physics and Applications. Nuclear Physics A, 752, 633-644. doi: 10.1016/j.nuclphysa.2005.02.132

Lorentz, H. A. (1952). The theory of electrons and its applications to the phenomena of light and radiant heat. New York: Dover Publications.

MIT OpenCourseWare | Free Online Course Materials. (n.d.). Retrieved July 03, 2014, from http://ocw.mit.edu/index.htm

McWilliam, A., & Rauch, M. (2004). Origin and evolution of the elements. Cambridge, UK: Cambridge University Press.

Mcmichael, R., & Stiles, M. (2009). A new connection between electricity and magnetism. Physics, 2. doi: 10.1103/Physics.2.11

Mercury Magazine. (n.d.). Retrieved July 05, 2014, from http://www.astrosociety.org/publications/mercury-magazine/

Moskowitz, C. (2013). Electron appears spherical, squashing hopes for new physics theories. Nature. doi: 10.1038/nature.2013.14162

Moulton, F. R., & Schifferes, J. J. (1960). The autobiography of science. Garden City, NY: Doubleday.

Myers, R. L. (2006). The basics of physics. Westport, CT: Greenwood Press.

Norton, O. R. (1994). Rocks from space: Meteorites and meteorite hunters. Missoula, MT: Mountain Press Pub.

Nuclear Physics B. (1990). Nuclear Physics B, 333(1), Iii. doi: 10.1016/0550-3213(90)90220-8

Daintith, J. (2009) Oxford Dictionary of Physics, 6th Ed. Oxford University Press

Panek, R. (2011). The 4 percent universe: Dark matter, dark energy, and the race to discover the rest of reality. Boston: Houghton Mifflin Harcourt.

Pasachoff, N. E. (1996). Marie Curie and the science of radioactivity. New York: Oxford University Press.

Pfeffer, J. I., & Nir, S. (2013). Modern physics: An introductory text. London: Imperial College Press.

Physics Outreach. (n.d.). Retrieved July 05, 2014, from http://www.aps.org/programs/outreach/

Pillai, S. O. (2010). Solid state physics. Kent: New Age Science.

Poincaré, L. (1907). The new physics and its evolution. London: K. Paul, Trench, Trübner, &.

Poundstone, W. (1999). Carl Sagan: A life in the cosmos. New York: Henry Holt.

Quantum Physics. (n.d.). Retrieved July 05, 2014, from http://quantumphysics.iop.org/

Randall, L. (n.d.). Higgs discovery: The power of empty space.

Recently on Physics Today online. (2010, 12). Physics Today, 63(10), 30. doi: 10.1063/1.3502546

Richtmyer, F. K. (1955). Introduction to modern physics. New York: McGraw-Hill.

Romer, R. H. (1999). Physics Education Research—A Supplement to the American Journal of Physics. American Journal of Physics, 67(6), 471. doi: 10.1119/1.19307

S., C. E. (1968). Editorial: Modern Physics—All Year Long. The Physics Teacher, 6(5), 263. doi: 10.1119/1.2352470

Schneider, H., & Wood, S. (2009). Backyard guide to the night sky. Washington, D.C.: National Geographic.

Science News, Articles, and Information - Scientific American. (n.d.). Retrieved July 05, 2014, from http://www.scientificamerican.com/

Scitation|Physics Today. (n.d.). Retrieved July 05, 2014, from http://scitation.aip.org/content/aip/magazine/physicstoday

Siegel, W. (1988). Introduction to string field theory. Singapore: World Scientific.

Simon, S. (1992). Our solar system. New York: Morrow Junior Books.

Singh, S. (2010). Big Bang. Rio de Janeiro (RJ): Record.

Slosson, E. E. (1920). Easy lessons in Einstein: A discussion of the more intelligible features of the theory of relativity. New York: Harcourt, Brace and Howe.

Smarandache, F., & Christianto, V. (2007). Hadron models and related new energy issues. United States: InfoLearnQuest Publisher.

Straka, J. M. (2009). Cloud and precipitation microphysics: Principles and parameterizations. Cambridge: Cambridge University Press.

Stratton, J., & Chu, L. (1939). Diffraction Theory of Electromagnetic Waves. Physical Review, 56(1), 99-107. doi: 10.1103/PhysRev.56.99

TIME. (n.d.). Retrieved July 03, 2014, from http://time.com/

Thompson, J. M. (2001). Visions of the future: Physics and electronics. New York: Cambridge University Press.

University of Oxford Department of Physics. (n.d.). Retrieved July 05, 2014, from http://www2.physics.ox.ac.uk/

Urone, P. P., Hinrichs, R., Dirks, K., & Sharma, M. (2013). College physics. Houston, TX: OpenStax College, Rice University.

Verma, H. C. (1999). Concepts of physics. New Delhi: Bharati Bhawan.

Viennot, L., & Collister, C. (2014). Thinking in physics: The pleasure of reasoning and understanding. Dordrecht: Springer.

Volovik, G. E. (2003). The universe in a helium droplet. Oxford: Clarendon Press.

Who is using HyperPhysics? (n.d.). Retrieved July 05, 2014, from http://hyperphysics.phy-astr.gsu.edu/hbase/hph.html

Wills, A. P. (1910). The Theory of Electrons and Its Applications to the Phenomena of Light and Radiant Heat. Science, 31(789), 221-223. doi: 10.1126/science.31.789.221-a

Yu, Z. (1999). A Possible Class of Two-Mode Photon States and their Properties. Modern Physics Letters B [Condensed Matter Physics; Statistical Physics and Applied Physics], 13(18), 639-644. doi: 10.1142/S0217984999000804

The free physics textbook. (n.d.). Retrieved July 05, 2014, from http://www.motionmountain.net/?gclid=CjkKEQjwxN6dBRDavZLpufuu9L8BEiQA4Vx1b-QeMflmNdAIASPOPS0C55rQ_8Jmpqnha_L6tQ9-05Dw_wcB

Physics.org | Home. (n.d.). Retrieved July 05, 2014, from http://physics.org/

Physicsworld.com. (n.d.). Retrieved July 05, 2014, from http://physicsworld.com/Index

Index

A

B

Mita Thakur

C

D

E

F

G

H

I

J

K

L

M

O

P

Q

R

Made in United States
Orlando, FL
04 November 2026

90204423R00070